Wavelets and Other Orthogonal Systems With Applications

Gilbert G. Walter

CRC Press

Boca Raton   Ann Arbor   London   Tokyo

Library of Congress Cataloging-in-Publication Data

Walter, Gilbert G.
    Wavelets and other orthogonal systems with applications / Gilbert G. Walter
       p. cm.
    Includes bibliographical references and index.
    ISBN 0-8493-7878-8
    1. Wavelets. I. Title.
QA403.3.W34   1994
515′.2433—dc20   94-13866
for Library of Congress   CIP

Preface

The subject of wavelets has evolved very rapidly in the last five or six years—so rapidly that many articles and books are already obsolete. However, there is one portion of wavelet theory that has reached a plateau, that is, the subject of orthogonal wavelets. The major concepts have become standard, and further development will probably be at the margins. In one sense they are no different than other orthogonal systems. They enable one to represent a function by a series of orthogonal functions. But there are notable differences: wavelet series converge pointwise when others don't, wavelet series are more localized and pick up edge effects better, wavelets use fewer coefficients to represent certain signals and images.

Unfortunately, not all is rosy. Wavelet expansions change excessively under arbitrary translations—much worse than Fourier Series. The same is true for other operators such as convolution and differentiation.

In this book wavelets are presented in the same setting as other orthogonal systems, in particular Fourier series and orthogonal polynomials. Thus their advantages and disadvantages can be seen more directly.

The level of the book is such that it should be accessible to engineering and mathematics graduate students. It will for the most part assume a knowledge of analysis at the level of beginning graduate real and complex analysis courses. However, some of the later chapters are more technical and will require a stronger background. The Lebesgue integral will be used throughout. This has no practical effect on the calculation of integrals but does have a number of theoretical advantages.

Wavelets constitute the latest addition to the subject of orthogonal series, which are motivated by their usefulness in applications. In fact, orthogonal series have been associated with applications from their inception. Fourier invented trigonometric Fourier series in order to solve the partial differential equation associated with heat conduction and wave propagation. Other orthogonal series involving polynomials appeared in the 19th century. These too were closely related to problems in partial differential equations. The Legendre polynomials are used to find solutions to Laplace's equation in the sphere, the Hermite polynomials and Laguerre polynomial for special cases of Schrödinger wave equations. These, together with Bessel functions, are special cases of Sturm-Liouville problems, which lead to orthogonal series, which are used to solve various partial differential equations.

The arrival of the Lebesgue integral in the early 20th century allowed the development of a general theory of orthogonal systems. While not oriented to applications, it allowed the introduction of new systems such as the Haar and Walsh systems, which have proven useful in signal processing. Also useful in this subject are the sinc functions and their translates, which form an orthogonal basis of a Paley-Wiener space. These are related to the prolate spheroidal functions, which are solutions both to an integral equation and a Sturm-Liouville problem.

The orthogonal sequences of wavelets, which are generalizations of the Haar system and the sinc system, have a number of unique properties. These make them useful in data compression, in image analysis, in signal processing, in numerical analysis, and in acoustics. They are particularly useful in digitizing data because of their decomposition and reconstruction algorithms. They also have better convergence properties than the classical orthogonal systems.

While the Lebesgue integral made a general theory of orthogonal systems possible, it is insufficiently general to handle many of the applications. In particular, the delta "function" or impulse function plays a central role in signal processing but is not a square integrable function. Fortunately a theory that incorporates such things appeared in the middle of the 20th century. This is the theory of "distributions," due mainly to L. Schwartz. It also is related to orthogonal systems in that it allows representation of distributions by orthogonal systems and also allows representations of functions by orthogonal distributions.

The body of the book is divided into 13 chapters of which the first 7 are expository and general while the remaining are more specialized and deal with applications to other areas. Each will be concerned with the use of or properties of orthogonal series.

In Chapter One we present two orthogonal systems that are prototypes for wavelets. These are the Haar system and the Shannon system, which have many, but not all, of the properties of orthogonal wavelets. They will be preceded by a section on general orthogonal systems. This is a standard theory that contains some results that will be useful in all of the particular examples.

Chapter Two will give a short introduction to tempered distributions. This is a relatively simple theory and is the only type of generalized function needed for much of orthogonal series. Many engineers still seem to apologize for their use of a "delta function". There is no need to do so since these are well defined proper mathematical entities. Included here also is the associated theory of Fourier transforms that enables one to take Fourier transforms of things like polynomial and trigonometric functions.

Chapter Three contains an introduction to the general theory of orthogonal wavelets. Their construction by a number of different schemes is given as are a number of their properties. These include their multiresolution property in which the terms of the series are naturally grouped at each resolution. The decomposition and reconstruction algorithms of Mallat, which give the coefficients at one resolution in terms of others, are presented here. Some of these properties are extended to tempered distributions in Chapter Five.

In Chapter Four we return to trigonometric Fourier series and discuss more detailed properties such as pointwise convergence and summability. These are fairly well known and many more details may be found in Zygmund's book. A short presentation on expansion of distributions in Fourier series is also presented.

In Chapter Five we also consider orthogonal systems in Sobolev spaces. These can be composed of delta functions as well as ordinary functions. In the former case we obtain an orthonormal series of delta function wavelets.

Chapter Six is devoted to another large class of examples, the orthogonal polynomials. The classical examples are defined and certain of their properties discussed. The Hermite polynomials are naturally associated with tempered distribution; properties of this connection are covered. Other orthogonal series are discussed in Chapter Seven.

Various kinds of convergence of orthogonal series are discussed in Chapter Eight. In particular, pointwise convergence of wavelet series is compared to that of other orthogonal systems. Also, the rate of convergence in Sobolev spaces is determined. Gibbs' phenomenon for wavelet series is compared to that for other series.

Chapter Nine deals with sampling theorems. These arise from many orthogonal systems including the trigonometric and polynomial systems. But the classical Shannon sampling theorem deals with wavelet subspaces for the Shannon wavelet. This can be extended to other wavelet subspaces as well. Both regular and irregular sampling points are considered.

In Chapter Ten we cover the relation between the translation operator and orthogonal systems. Wavelet expansions are not very well behaved with respect to this operator except for certain examples.

Chapter Eleven deals with analytic representation based on both Fourier series and wavelets. These are used to solve boundary value problems for harmonic functions in a half-plane with specified values on the real line.

Chapter Twelve covers probability density estimation with various orthogonal systems. Both Fourier series and Hermite series have been used, but wavelets come out the best.

Finally in the last chapter we cover the Karhunen-Loève theory for representing stochastic processes in terms of orthogonal series. An alternate formulation based on wavelets is developed.

Some of this text material was presented to a graduate course of mixed mathematics and engineering students. While not directly written as a text, it can serve as the basis for a modern course in Special Functions or in mathematics of signal processing. Problems are included at the end of each chapter. For the most part these are designed to aid in the understanding of the text material.

Acknowledgements

Many persons helped in the preparation of the manuscript for this book, but two deserve special mention: Joyce Miezin for her efficient typing and ability to convert my handwriting into the correct symbols, and Bruce O'Neill for catching many of my mathematical misprints.

G. G. Walter

Table of Contents

1

Orthogonal Series

Orthogonal series play an important part in many areas of mathematics as well as in applications. They constitute an easy way of representing a function in terms of a series and may replace complicated operators on the function by simpler ones on the coefficients of the series. The most familiar orthogonal systems are the trigonometric and the various orthogonal polynomials. Not so familiar but becoming increasingly widely used are the Haar, the Shannon, and wavelet systems.

The basic theory of orthogonal series is deceptively simple, but its detailed study contains many surprisingly difficult questions. In this chapter we skip the latter and present only a few elements of the theory. We first present a little of the general theory and then discuss a few of the principal examples. One, the trigonometric system, will be an important tool in subsequent chapters. The other two, the Haar and the Shannon systems, will serve as prototypes for the construction of wavelets.

1.1   General theory

While there exist many different orthogonal systems, they all have a number of properties in common, which we present here. We shall restrict ourselves to $L^2(a, b)$; the set of square (Lebesgue) integrable functions on (a, b), a real interval. The theory is the same if we introduce a weight function or even if we consider general separable Hilbert spaces.

A nontrivial sequence $\{f_n\}_{n=0}^{\infty}$ of real (or complex) functions in $L^2(a, b)$ is said to be <u>orthogonal</u> if

$$\langle f_n, f_m \rangle = \int_a^b f_n(x)\overline{f_m(x)}\, dx = 0, \qquad n \neq m, \;\; n, m = 0, 1, 2, \ldots$$

and <u>orthonormal</u> if in addition $\langle f_n, f_n \rangle = 1$, $n = 0, 1, 2, \cdots$. For example, $f_n(x) = \sin(n+1)x$ is orthogonal on $(0, \pi)$. Another example is

$$f_n(x) = \chi_{[n,n+1)}(x) = \begin{cases} 1, & n \le x < n+1 \\ 0, & 0 \le x < n, \ n+1 \le x \end{cases}$$

which in fact is orthonormal on $[0, \infty)$.

The idea is to expand a given function $f(x) \in L^2(a, b)$ in an orthonormal series

$$f(x) = \sum_{n=0}^{\infty} c_n f_n(x). \tag{1.1}$$

This is not always possible (e.g., take $f(x) = \chi_{[0.5,1)}(x)$ in the second example), but if it is, then the c_n's must have a special form. We shall use the usual notation for the L^2 norm, $\|f\| = \langle f, f \rangle^{1/2}$.

PROPOSITION 1.1
Let $\{c_n\}$ be a sequence such that the series in (1.1) converges in the sense of $L^2(a, b)$ to $f(x)$; then $c_n = \langle f, f_n \rangle$.

The proof is immediate. We multiply both sides of (1.1) by $\overline{f_m(x)}$ and then integrate. Because of the orthogonality all the terms in the series drop out except c_m. There is no problem with interchanging the integral and the summation because of the continuity of the inner product with respect to the norm.

Convergence in the sense of $L^2(a, b)$ is also known as <u>mean square convergence</u>, and the error

$$e_N = \left\| f - \sum_{n=0}^{N} c_n f_n \right\|^2$$

is called the <u>mean square error</u>. The coefficients appearing in Proposition 1.1 are called the <u>Fourier coefficients</u> of f with respect to $\{f_n\}$ and have another property that makes them useful.

PROPOSITION 1.2
Let $\{c_n\}$ be the Fourier coefficient of $f \in L^2(a, b)$ and $\{a_n\}$ any other sequence; then we have

$$\left\| f - \sum_{n=0}^{N} c_n f_n \right\|^2 \le \left\| f - \sum_{n=0}^{N} a_n f_n \right\|^2,$$

i.e., the mean square error is minimized for the series with Fourier coefficients.

The proof is obtained by adding and subtracting the series with the c_n's:

$$\left\| f - \sum_{n=0}^{N} a_n f_n \right\|^2 = \langle f - \sum_{n=0}^{N} a_n f_n, \ f - \sum_{n=0}^{N} a_n f_n \rangle$$

$$= \langle f, f \rangle - \sum a_n \langle f_n, f \rangle - \sum \overline{a_n} \langle f, f_n \rangle + \sum |a_n|^2$$

$$= \langle f, f \rangle - \sum (a_n \overline{c_n} + \overline{a_n} c_n)$$

$$+ \sum |a_n|^2 + \sum |c_n|^2 - \sum |c_n|^2$$

$$= \langle f, f \rangle + \sum |a_n - c_n|^2 - \sum |c_n|^2$$

$$= \left\| f - \sum_{n=0}^{N} c_n f_n \right\|^2 + \sum_{n=0}^{N} |a_n - c_n|^2. \qquad (1.2)$$

Since $\sum |a_n - c_n|^2 \geq 0$, the conclusion follows. □

Another way of thinking of this is that the Fourier coefficients give the orthogonal projection of f onto the subspace V_N spanned by $(f_0, f_1, f_2, \ldots, f_N)$. Indeed by another simple calculation we see that

$$\langle \sum_{n=0}^{N} c_n f_n, \ f - \sum_{n=0}^{N} c_n f_n \rangle = 0.$$

The best approximation to f in V_N is given by this sum, and the error is orthogonal to V_N.

Similar calculations lead to <u>Bessel's inequality</u>

$$\sum_{n=0}^{\infty} |c_n|^2 \leq \|f\|^2 \qquad (1.3)$$

since

$$0 \leq \left\| f - \sum_{n=0}^{N} c_n f_n \right\|^2 = \|f\|^2 - \sum_{n=0}^{N} |c_n|^2,$$

and therefore $\left\{ \sum_{n=0}^{N} |c_n|^2 \right\}$ is a monotone sequence bounded by $\|f\|^2$. Thus the series of (1.3) converges and has the same bound. A simple consequence of Bessel's inequality is that $\{c_n\} \in \ell^2$ and $c_n \to 0$ as $n \to \infty$.

To round out our theory, we should like to have the series with Fourier coefficients $\sum c_n f_n$ converge to f. By Bessel's inequality the partial sums are a Cauchy sequence in $L^2(\mathbb{R})$, which because of the completeness of this space must converge in the L^2 sense but not to f necessarily. To ensure this we need to add another condition, the <u>completeness</u> of the system. The orthonormal system $\{f_n\}$ is said to be <u>complete</u> in $L^2(a,b)$ if no nontrivial $f \in L^2(a,b)$ is orthogonal to all the f_n's, i.e., if $\langle f, f_n \rangle = 0$, $n = 0, 1, 3, \ldots$, for $f \in L^2(a,b)$ then $f = 0$, a.e.

THEOREM 1.3
Let $\{f_n\}$ be an orthonormal system in $L^2(a,b)$; let $f \in L^2(a,b)$ with Fourier coefficients $\{c_n\}$; then

$$\left\| f - \sum_{n=0}^{N} c_n f_n \right\| \to 0 \qquad as\, N \to \infty$$

if and only if $\{f_n\}$ is complete.

PROOF By Bessel's inequality we know that the series (1.1) converge to some $g \in L^2(a,b)$, and hence

$$f - \sum_{n=0}^{N} c_n f_n \to f - g\,.$$

Now the Fourier coefficients of g are given by

$$\langle g, f_m \rangle = \lim_{N \to \infty} \langle \sum_{n=0}^{N} c_n f_n, f_m \rangle = c_m$$

and hence are the same as those of f. Thus $f - g$ has all zero coefficients, and, if the system is complete, $f - g = 0$ a.e. Since the series converges to g it must also converge to f.

On the other hand if the series converges to f, and all the coefficients are zero, then $f = 0$ a.e. as well.   □

The conclusion of the theorems can be restated as Parseval's equality

$$\|f\|^2 = \sum_{n=0}^{\infty} |c_n|^2 \qquad\qquad (1.4)$$

since

$$\left\| f - \sum_{n=0}^{N} c_n f_n \right\|^2 = \|f\|^2 - \sum_{n=0}^{N} |c_n|^2\,.$$

An alternate form is given by

$$\sum_{k=0}^{\infty} c_k \overline{d_k} = \langle f, g \rangle$$

where $d_k = \langle g, f_k \rangle$. This is obtained by applying (1.4) to $f + g$ and $f - g$ and then subtracting.

Either of these results can be taken as test for completeness. However, it is sufficient to check (1.4) for a set of functions $\{h_k\}$ whose closed linear span (i.e., the closure in the sense of $L^2(a, b)$ of $\sum_{k=0}^{K} a_k h_k$) is $L^2(a, b)$. In particular we may take h_k to be the set of characteristic functions of intervals.

1.2 Examples

There are many examples of orthonormal systems in the mathematical literature (see [Al], [O], [Sc]). The earliest and most widely studied is the trigonometric system which we consider in more detail below. We also consider two more recent systems, the Haar and the Shannon, which form prototypes of the newest system, the orthogonal wavelets. In a later chapter we study some aspects of orthogonal polynomials. But there are many others, e.g., Sturm-Liouville systems, which are used to solve partial differential equations; the Walsh functions, which are piecewise constant; and the eigenfunctions of a compact symmetric integral operator, which will not be covered in detail.

1.2.1 Trigonometric system

The trigonometric system is a complete orthogonal system in $L^2(-\pi, \pi)$ given by

$$f_0(x) = 1/2, \quad f_1(x) = \sin x, \quad f_2(x) = \cos x,$$
$$\cdots f_{2n-1}(x) = \sin nx, \quad f_{2n}(x) = \cos nx, \cdots.$$

It is usually not normalized, since $\|f_n\|^2 = \pi$, $n \neq 0$. The series is usually written in the form

$$S(x) = \frac{a_o}{2} + \sum_{n=0}^{\infty} a_n \cos nx + b_n \sin nx . \tag{1.5}$$

If (1.5) is the Fourier series of a function $f \in L^2(-\pi, \pi)$, the coefficients are given by

$$a_n = \frac{1}{\pi} \int_{-\pi}^{\pi} f(x) \cos nx \, dx, \qquad n = 0, 1, \cdots$$
$$b_n = \frac{1}{\pi} \int_{-\pi}^{\pi} f(x) \sin nx \, dx, \qquad n = 1, 2, \cdots . \tag{1.6}$$

The orthogonality of this system is easy to prove by using a few trigonometric identities. However, the completeness, though well known, is not so obvious. In the interest of completeness we present a proof. It involves first showing that the Fourier series converges uniformly for certain functions. It should be remarked that this is not true for all continuous functions; there are examples where the Fourier series fails to converge on a dense set of points [Z, p. 298].

If the series (1.5) is to converge uniformly, then the limit function must be continuous and periodic of period 2π, which we assume f to be. We shall need an expression for the partial sums of the series, obtained by substituting (1.6) into the partial sums of (1.5).

$$
\begin{aligned}
S_n(x) &= \frac{a_0}{2} + \sum_{k=1}^{n} a_k \cos kx + b_k \sin kx \\
&= \frac{1}{\pi} \int_{-\pi}^{\pi} f(t) \left\{ \frac{1}{2} + \sum_{k=1}^{n} \cos kt \cos kx + \sin kt \sin kx \right\} dt \\
&= \frac{1}{\pi} \int_{-\pi}^{\pi} f(t) \left\{ \frac{1}{2} + \sum_{k=1}^{n} \cos k(x-t) \right\} dt \qquad (1.7) \\
&= \frac{1}{\pi} \int_{-\pi}^{\pi} f(t) \frac{\sin\left(n+\frac{1}{2}\right)(x-t)}{2\sin(x-t)/2} dt \\
&= \int_{-\pi}^{\pi} f(x-u) \frac{\sin\left(n+\frac{1}{2}\right)u}{2\pi \sin u/2} du \,.
\end{aligned}
$$

The expression

$$
D_n(u) = \frac{1}{\pi} \left[\frac{1}{2} + \sum_{k=1}^{n} \cos ku \right] = \frac{\sin\left(n+\frac{1}{2}\right)u}{2\pi \sin u/2} \qquad (1.8)
$$

is called the Dirichlet kernel, and it plays a central role in the study of pointwise convergence of Fourier series. It may be shown true by multiplying both sides of

$$
\pi D_n(u) = \frac{1}{2} + \sum_{k=1}^{n} \cos ku
$$

by $\sin \frac{u}{2}$ and then forming a telescoping sum.

PROPOSITION 1.4
Let f be a 2π periodic function in $C^2(I\!R)$; then

$$\text{(i)} \quad \sup_{x \in I\!R} |S_n(x) - f(x)| \to 0$$

$$\text{(ii)} \quad \|S_n - f\| \to 0$$

as $n \to \infty$.

PROOF   Since $\int_{-\pi}^{\pi} D_n(u)du = 1$, the difference between S_n and f may be expressed as

$$
\begin{aligned}
S_n(x) - f(x) &= \int_{-\pi}^{\pi} \{f(x-u) - f(x)\} D_n(u) du \\
&= \frac{1}{\pi} \int_{-\pi}^{\pi} \left\{ \frac{f(x-u) - f(x)}{2\sin u/2} \right\} \sin\left(n + \frac{1}{2}\right) u \, du \\
&= \frac{1}{\pi} \int_{-\pi}^{\pi} g(x,u) \sin\left(n + \frac{1}{2}\right) u \, du \qquad (1.9) \\
&= -\frac{1}{\pi} \frac{\cos\left(n + \frac{1}{2}\right) u}{n + \frac{1}{2}} g(x,u) \Big|_{-\pi}^{\pi} \\
&\quad + \frac{1}{\pi} \int_{-\pi}^{\pi} \frac{\partial g(x,u)}{\partial u} \frac{\cos\left(n + \frac{1}{2}\right) u}{\left(n + \frac{1}{2}\right)} du \,,
\end{aligned}
$$

where $g(x,u) = \frac{f(x-u)-f(x)}{2\sin u/2}$.

Since both $g(x,u)$ and its derivative are uniformly bounded, the last expression gives us

$$|S_n(x) - f(x)| \leq \frac{C}{\left(n + \frac{1}{2}\right)}$$

for some constant C. Hence (i) must hold and by squaring and integration of (i), so must (ii).   □

Since twice differentiable functions are dense in $L^2(-\pi, \pi)$, it follows that the trigonometric system is complete.

An alternative form for the trigonometric series (1.5) is the exponential form

$$S(x) = \sum_{n=-\infty}^{\infty} c_n e^{inx} \,, \qquad (1.10)$$

where convergence is with respect to the symmetric partial sums. If (1.10) is a Fourier series the coefficients are

$$c_n = \frac{1}{2\pi} \int_{-\pi}^{\pi} f(x) e^{-inx} \, dx \,. \qquad (1.11)$$

Of course expression (1.10) is reducible to (1.5) by using $e^{\pm inx} = \cos nx \pm i \sin nx$.

Another way of looking at these expressions is as transforms.

The <u>finite Fourier transform</u> of the periodic function $f(x)$ is given by (1.11) with the inverse transform given by (1.10). It converts a differential operator into a multiplication operator,

$$\frac{1}{2\pi} \int_{-\pi}^{\pi} \left(Df(x)\right) e^{-inx}\, dx = in\, c_n\, ,$$

which makes it a useful tool in differential equations. Its absolute value is also shift invariant,

$$\left| \frac{1}{2\pi} \int_{-\pi}^{\pi} f(x - \alpha) e^{-inx}\, dx \right| = |c_n|\, .$$

The <u>infinite Fourier transform</u> of the function $f \in L^1(-\infty, \infty)$ is the expression

$$\hat{f}(w) = \int_{-\infty}^{\infty} f(t) e^{-iwt}\, dt\, , \qquad w \in \mathbb{R}\, . \tag{1.12}$$

The image of the transform in this case is a continuous function on $\mathbb{R}$, which, if it is also in $L^1(\mathbb{R})$, leads to the inverse [B-C, p. 19]

$$f(t) = \frac{1}{2\pi} \int_{-\infty}^{\infty} \hat{f}(w) e^{iwt}\, dw\, , \qquad t \in \mathbb{R}\, . \tag{1.13}$$

Versions of Parseval's equality (1.4) also exist for Fourier transforms. They are for $f, g \in L^2$ [B-C, p. 105]

$$\|f\|^2 = \frac{1}{2\pi} \|\hat{f}\|^2 \qquad \langle f, g \rangle = \frac{1}{2\pi} \langle \hat{f}, \hat{g} \rangle\, .$$

This requires a more general definition of Fourier transform, which together with other properties are found in the next chapter.

One can also go the other way and approximate (1.11) by a discrete sum. This gives us the <u>discrete Fourier transform</u>,

$$\gamma_k = \frac{1}{N} \sum_{j=0}^{N-1} f\left(\frac{2\pi j}{N}\right) e^{-ijk2\pi/N}\, , \qquad k = 0, \cdots, N-1\, , \tag{1.14}$$

with the inverse given by

$$f\left(\frac{2\pi j}{N}\right) = \sum_{k=0}^{N-1} \gamma_k\, e^{ikj2\pi/N}\, , \qquad j = 0, \cdots, N-1\, . \tag{1.15}$$

This is the form that leads to the <u>fast Fourier transform</u> which, by grouping terms in (1.14), reduces the computation time considerably. This has made transform methods much more useful in partial differential equations, image processing, time series, and other applied problems [S, p. 448].

1.2.2 Haar system

The Haar orthogonal system begins with $\phi(t)$, the characteristic function of the unit interval

$$\phi(t) = \chi_{[0,1)}(t) \, .$$

It is clear that $\phi(t)$ and $\phi(t-n)$, $n \neq 0$, $n \in \mathbb{Z}$ are orthogonal since their product is zero. It is also clear that $\{\phi(t-n)\}$ is not a complete orthogonal system in $L^2(\mathbb{R})$ since its closed linear system V_0 consists of piecewise constant functions with possible jumps only at the integers. The characteristic function of $[0, 1/2)$, for example, cannot have a convergent expansion.

In order to include more functions we consider the dilated version of $\phi(t)$ as well, $\phi(2^m t)$ where $m \in \mathbb{Z}$. Then by a change of variable we see that $\{2^{m/2}\phi(2^m t - n)\}$ is an orthonormal system. Its closed linear span will be denoted by V_m. Since any function in $L^2(\mathbb{R})$ may be approximated by a piecewise constant function f_m with jumps at binary rationals, it follows that $\bigcup_m V_m$ is dense in $L^2(\mathbb{R})$. Thus the system $\{\phi_{mn}\}$ where

$$\phi_{mn}(t) = 2^{m/2}\phi(2^m t - n)$$

is complete in $L^2(\mathbb{R})$, but, since $\phi(t)$ and $\phi(2t)$ are not orthogonal, it is not an orthogonal system.

Fortunately the cure is simple; we let $\psi(t) = \phi(2t) - \phi(2t - 1)$. Then everything works; $\{\psi(t-n)\}$ is an orthonormal system, and $\psi(2t-k)$ and $\psi(t-n)$ are orthogonal for all k and n. This enables us to deduce that $\{\psi_{mn}\}_{m,n\in\mathbb{Z}}$ where

$$\psi_{mn}(t) = 2^{m/2}\psi(2^m t - n)$$

is a complete orthonormal system in $L^2(\mathbb{R})$. This is the Haar system; the expansion of $f \in L^2(\mathbb{R})$ is

$$f(t) = \sum_{m=-\infty}^{\infty} \sum_{n=-\infty}^{\infty} \langle f, \psi_{mn}\rangle \psi_{mn}(t) \, , \qquad (1.16)$$

with convergence in the sense of $L^2(\mathbb{R})$. The standard approximation is the series given by

$$f_m(t) = \sum_{k=-\infty}^{m-1} \sum_{n=-\infty}^{\infty} \langle f, \psi_{kn}\rangle \psi_{kn}(t) \, . \qquad (1.17)$$

It converges to a piecewise constant function with jumps at $2^{-m}n$, $n \in \mathbb{Z}$, at most.

Hence, $f_m \in V_m$ and since, by Parseval's equality,

$$\langle f_m, \phi_{mn} \rangle = \sum_{k=-\infty}^{m-1} \sum_{j=-\infty}^{\infty} \langle f, \psi_{kj} \rangle \langle \psi_{kj}, \phi_{mn} \rangle = \langle f, \phi_{mn} \rangle,$$

it is the projection of f on V_m, i.e., $f_m = \sum_n \langle f, \phi_{mn} \rangle \phi_{mn}$.

This enables us to get a pointwise convergence theorem.

PROPOSITION 1.5
Let f be continuous on $\mathbb{R}$ and have compact support; then $f_m \to f$ uniformly.

PROOF Since f is uniformly continuous, it follows that for each $\epsilon > 0$ there exists an m such that

$$|f(x) - f(y)| < \epsilon \qquad \text{when } |x - y| \leq 2^{-m}.$$

For $x \in [n2^{-m}, (n+1)2^{-m})$ we have

$$f_m(x) = 2^{m/2} \int_{2^{-m}n}^{2^{-m}(n+1)} f(t)dt \, 2^{m/2} \phi(2^m x - n)$$

since all the other terms in the series are zero in this interval. Therefore by the mean value theorem

$$f_m(x) = f(\zeta) 2^{-m} 2^m \phi(2^m x - n) = f(\zeta),$$

and since $|x - \zeta| \leq 2^{-m}$, $|f_m(x) - f(x)| < \epsilon$. $\square$

This Haar system is our first prototype of a wavelet system, and we shall return to it several times later. At this point it should be observed that the uniform convergence of f_m to f is a property not shared by the trigonometric system. The uniform convergence of (1.16) also follows since the inner series has only a finite number of terms, and the partial sums of the outer series converges to zero uniformly since they are of the form $f_m(x) - f_{-p}(x)$.

The $\phi(t)$ is usually called the scaling function in wavelet terminology while $\psi(t)$ is the mother wavelet.

1.2.3 The Shannon system

A second prototype also begins with the characteristic function of an interval. Now, however, it is the Fourier transform of the scaling function, taken to be

$$\hat{\phi}(w) = \begin{cases} 1 & -\pi \leq w < \pi \\ 0 & \text{o.w.} \end{cases}.$$

Its inverse Fourier transform is

$$\phi(t) = \frac{1}{2\pi} \int_{-\infty}^{\infty} \hat{\phi}(w)e^{iwt}dw = \frac{1}{2\pi} \int_{-\pi}^{\pi} e^{iwt}dw = \frac{\sin \pi t}{\pi t}.$$

The orthogonality of $\phi(t)$ and $\phi(t-n)$ is based on properties of the Fourier transform, of Parseval's equality, and the fact that $(\widehat{\phi(t-\alpha)})(w) = \hat{\phi}(w)e^{-i\alpha w}$:

$$\int_{-\infty}^{\infty} \phi(t)\phi(t-n)dt = \frac{1}{2\pi} \int_{-\infty}^{\infty} \hat{\phi}(w)\overline{\hat{\phi}(w)}e^{iwn}dw$$

$$= \frac{1}{2\pi} \int_{-\pi}^{\pi} e^{iwn}dw = \frac{\sin \pi n}{\pi n} = 0, \, n \neq 0.$$

Let $f(t)$ be a function that is square integrable and whose Fourier transform $\hat{f}(w)$ vanishes for $|w| > \pi$. It has a Fourier series given by

$$\hat{f}(w) = \sum_n c_n e^{iwn}, \qquad |w| \leq \pi \tag{1.18}$$

where $c_n = \frac{1}{2\pi} \int_{-\pi}^{\pi} \hat{f}(w)e^{-iwn}dw$. By the Fourier integral thereom (1.13) this is just $f(-n)$. This theorem applied to both sides of (1.18) yields

$$f(t) = \frac{1}{2\pi} \int_{-\pi}^{\pi} \hat{f}(w)e^{iwt}dw$$

$$= \sum_n f(-n)\frac{1}{2\pi} \int_{-\pi}^{\pi} e^{iwn}e^{iwt}dw \tag{1.19}$$

$$= \sum_n f(-n)\frac{\sin \pi(t+n)}{n(t+n)}.$$

We denote by V_0 the set of all such functions. This is a linear space and is closed as well since limits (in the square integrable sense) of the sequences of functions in V_0 are also in V_0.

The formula (1.19) is referred to as the <u>Shannon sampling theorem</u>. It enables one to recover a band-limited function in V_0 from its values on the integers. This is used by engineers to convert a digital to an analog signal (as in compact discs). See [B] and [Za] for more properties.

By changing the scale in (1.19) by a factor of 2, we can obtain a sampling theorem on the half integers. (Let $2x = t$ and let $g(x) = f(2x)$). The space with the new scale is V_1 and it will consist of functions whose Fourier transforms vanish outside of $[-2\pi, 2\pi]$. We may repeat this as often as we want and get thereby an increasing sequence of spaces.

We can also stretch the scale instead of shrinking it to obtain a sequence $\{V_m\}_{m=-\infty}^{\infty}$ satisfying

$$\cdots \subseteq V_{-m} \subseteq \cdots \subseteq V_{-1} \subseteq V_0 \subseteq V_1 \subseteq \cdots \subseteq V_m \subseteq \cdots .$$

As we go to the left in this sequence the support of the Fourier transform (the set outside of which it vanishes) shrinks to 0. As we go to the right, it expands to all of $\mathbb{R}$. Thus we have

(i) $\bigcap_{n} V_m = \{0\}$

and

(ii)   each $f \in L^2(\mathbb{R})$ can be approximated by a function in V_m for m sufficiently large.

The sequence $\{V_m\}$ is called a "multiresolution analysis" associated with $\phi(t)$.

Just as with the other prototype, we can introduce a function $\psi(t)$ in V_1 which is orthogonal to $\phi(t - n)$, $\psi(t) = 2\phi(2t) - \phi(t)$. Its Fourier transform is given by

$$\hat{\psi}(w) = \hat{\phi}\left(\frac{w}{2}\right) - \hat{\phi}(w)$$

and has support on $[-2\pi, -\pi] \cup [\pi, 2\pi]$. Since the supports of $\hat{\phi}$ and $\hat{\psi}$ are disjoint, the needed orthogonality follows. Thus, the inverse Fourier transform $\psi(t)$ is another example of a "mother wavelet", and leads to another orthonormal sequence $\{\psi_{m,n}(t)\}$. As in the case of the Haar system, it is also complete and the expansion of certain continuous functions converges uniformly. Rather than require this function to have compact support, we assume only that its Fourier transform $\hat{f} \in L^1(\mathbb{R})$.

PROPOSITION 1.6
Let $f \in L^2(\mathbb{R})$ have a Fourier transform $\hat{f} \in L^1(\mathbb{R})$; then $f_m \to f$ uniformly on $\mathbb{R}$.

PROOF   If $f_m = \sum_{n} \langle f, \phi_{mn} \rangle \phi_{mn}$, then

$$\hat{f}_m(w) = \sum_{n} \frac{1}{2^{m+1}\pi} \int_{-2^m\pi}^{2^m\pi} \hat{f}(\zeta)\hat{\phi}(2^{-m}\zeta)e^{i\zeta n2^{-m}} d\zeta \, \hat{\phi}(2^{-m}w)e^{-iwn2^{-m}}.$$

The right hand side is the Fourier series of $\hat{f}(w)$ on $[-2^m\pi, 2^m\pi]$ and hence is equal to the restriction of $\hat{f}(w)$ to this interval,

$$\hat{f}_m(w) = \hat{f}(w)\chi_{2^m\pi}(w) \tag{1.20}$$

where $\chi_{2^m\pi}(w)$ is the characteristic function of $[-2^m\pi, 2^m\pi]$. The error then may be given by

$$f(t) - f_m(t) = \frac{1}{2\pi} \int_{-\infty}^{\infty} \left(\hat{f}(w) - \hat{f}_m(w)\right) e^{iwt} dw$$

$$= \frac{1}{2\pi} \left\{ \int_{2^m\pi}^{\infty} + \int_{-\infty}^{-2^m\pi} \right\} \hat{f}(w) e^{iwt} dw.$$

Hence we obtain

$$|f(t) - f_m(t)| \leq \frac{1}{2\pi} \left\{ \int_{2^m\pi}^{\infty} + \int_{-\infty}^{-2^m\pi} \right\} |\hat{f}(w)| \, dw$$

which, since $\hat{f} \in L^1(\mathbb{R})$, must converge to 0 as $m \to \infty$. $\quad\square$

The astute reader will recognize V_0 as the Paley-Wiener space of π band limited functions [P-W]. It is a reproducing kernel Hilbert space with reproducing kernel $q(x,t) = \frac{\sin \pi(x-t)}{\pi(x-t)}$. The expansion of $f \in V_0$ with respect to $\phi(x-n)$ given by (1.19) not only converges in $L^2(\mathbb{R})$ but converges uniformly on all of $\mathbb{R}$. The proof is implicit in (1.19).

1.3 Problems

1. The following are orthogonal systems in $L^2(0, \pi)$:

    (a) $\{\cos nx\}_{n=0}^{\infty}$,

    (b) $\{\sin nx\}_{n-1}^{\infty}$,

    (c) $\{\sin(n + \frac{1}{2})x\}_{n=0}^{\infty}$.

   Show they are also complete in $L^2(0, \pi)$.

2. Show that any orthogonal system $\{f_n\}$ is also linearly independent.

3. Let $\{f_n\}_{n=0}^{\infty}$ be a linearly independent sequence in $L^2(a, b)$; define $\phi_0 = f_0$, $\phi_1 = f_1 - \frac{\langle f_1, \phi_0 \rangle \phi_0}{\|\phi_0\|^2}$, $\phi_2 = f_2 - \frac{\langle f_2, \phi_1 \rangle \phi_1}{\|\phi_1\|^2} - \frac{\langle f_2, \phi_0 \rangle \phi_0}{\|\phi_0\|^2}, \cdots$. Show that $\{\phi_n\}$ is an orthogonal system.

4. Let $\{f_n\}$ be linearly independent and complete in $L^2(a, b)$ (i.e., the closed linear span of $\{f_n\}$ is $L^2(a, b)$). Then $\{f_n\}$ is said to be a Riesz basis if there are positive constants A and B such that

$$A \sum_{i=1}^{n} |c_i|^2 \leq \left\| \sum_{i=1}^{n} c_i f_i \right\|^2 \leq B \sum_{i=1}^{n} |c_i|^2.$$

Define $\{g_n\}$ to be the biorthogonal sequence to $\{f_n\}$ ($\langle g_n, f_k \rangle = \delta_{nk}$). Show that

(a) Each g_n may be extended to a continuous linear functional on $L^2(a, b)$ by defining $\langle g_n, f \rangle$ appropriately.

(b) For each $f \in L^2(a, b)$

$$f = \sum_{n=0}^{\infty} \langle f, g_n \rangle f_n.$$

(c) If $\{c_n\} \in \ell^2$ then $\sum_n c_n f_n$ converges in $L^2(a, b)$.

5. Let $f_1(x) = x$ and $f_2(x) = \pi^2 - 3x^2$, $-\pi < x < \pi$. Find the Fourier series of f_1 and f_2 on $[-\pi, \pi]$ and use them to sum the series

(a) $\sum_n \frac{(-1)^n}{n}$

(b) $\sum \frac{1}{n^2}$

(c) $\sum \frac{1}{n^4}$

(d) $\sum \frac{1}{n^6}$

6. Show that the Haar system

$$\{\phi(t),\ 2^{m/2}\psi(2^m t - n)\}, \quad m = 0, 1, 2, \cdots, \quad n = 0, 1, \cdots 2^m - 1,$$

is an orthonormal basis of $L^2(0, 1)$.

7. Find the expansion with respect to $\{\phi(t - n)\}$ the Shannon system in V_0 of the functions given by

(a) $f_1(w) = \frac{1}{2\pi} \int\limits_{-\pi}^{\pi} w\, e^{ixw}\, dw$

(b) $f_2(w) = \frac{1}{2\pi} \int\limits_{-\pi}^{\pi} (\pi^2 - 3w^2)\, dw.$

2

A Primer on Tempered Distributions

For many applications it is necessary to work with a larger class of functions than $L^1(\mathbb{R})$. It is also desirable to have a more general theory of the Fourier transform, which would enable us to calculate Fourier transforms of periodic functions or polynomials. This is impossible with the L^1 (or the usual L^2) theory, which does not even work for constant functions.

The simplest theory that extends the Fourier transform to all functions of polynomial growth is the theory of "tempered distributions". These include, in addition to functions, certain objects such as the "delta function," which is not strictly a function. Nonetheless, it is widely used and is central to some applications such as the study of transfer functions of linear filters in engineering.

These tempered distributions may be considered limits in some sense of sequences of ordinary $L^1(\mathbb{R})$ functions. In fact, one approach is to define them as limits of Cauchy sequences of such functions. In this case the familiar process of extending the rational numbers to obtain the real numbers is imitated (see [K]). However, there are other approaches that can be used for the formal definitions, but they will still be limits of sequences of functions and are best thought of as such.

In our approach, which is due to L. Schwartz [S], these ideal elements are defined as continuous linear functionals on a space of "test functions". The resulting objects will include all locally integrable functions of polynomial growth and all tempered measures as well as their derivatives of every order. They are closed under the Fourier transform and provide an appropriate setting for much of signal processing.

2.1   Tempered distributions

To define tempered distributions, we begin with the space $\mathcal{S}$ of rapidly decreasing C^∞ functions on $\mathbb{R}$, i.e., functions that satisfy

$$\left|\theta^{(k)}(t)\right| \le C_{pk}(1+|t|)^{-p}, \quad p,k = 0,1,2,\ldots, t \in \mathbb{R}. \qquad (2.1)$$

$\mathcal{S}$ is clearly a linear space and has an associated convergence given by the semi-norms

$$\gamma_{pk} = \sup_t (1+|t|)^p \left|\theta^{(k)}(t)\right|, \quad p,k = 0,1,2,\ldots. \qquad (2.2)$$

This means that $\theta_\nu \to \theta$ in $\mathcal{S}$ whenever

$$(1+|t|)^p D^k(\theta_\nu(t) - \theta(t)) \to 0$$

uniformly in t for each p and k as $\nu \to \infty$. (D is the derivative operator.)
    The space $\mathcal{S}$ contains the Hermite functions $\{h_n\}$ given by

$$h_o(x) = \frac{e^{-x^2/2}}{\sqrt{\pi}}, \qquad x \in \mathbb{R}$$

and recursively

$$(x - D)h_n(x) = \sqrt{2n+2}h_{n+1}(x), \quad n = 0,1,\ldots, x \in \mathbb{R}.$$

These constitute an orthonormal basis of $L^2(\mathbb{R})$ (see Chapter 6). Hence $\mathcal{S}$ must be dense in $L^2(\mathbb{R})$ in the L^2 norm. The C^∞ functions of compact support are also contained in $\mathcal{S}$.
    This space is complete with respect to the convergence of (2.2). The derivative operator is continuous with respect to this convergence, i.e., if $\theta_\nu \to \theta$ in the sense of (2.2) then $D\theta_\nu \to D\theta$ in the same sense. The same is true for the multiplication operator $M\theta(t) = F(t)\theta(t)$ where $F(t)$ is any function of polynomial growth in $C^\infty(\mathbb{R})$ (in particular, a polynomial).
    Two other operators that are important for wavelets are the dilation operator D_α, $\alpha > 0$ ($D_\alpha f(t) = f(\alpha t)$) and translation operator T_β ($T_\beta f(t) = f(t - \beta)$). Our space $\mathcal{S}$ is closed under these operators and both are continuous with respect to (2.2).
    The Fourier transform $\mathcal{F}$ given by

$$(\mathcal{F}\theta)(w) = \hat{\theta}(w) = \int_{-\infty}^{\infty} e^{-iwt}\theta(t)dt \qquad (2.3)$$

maps S into S. This follows from the fact that $\mathcal{F}$ converts differentiation into multiplication by a multiple of w and vice versa. In fact by the Fourier integral theorem

$$\theta(t) = \frac{1}{2\pi} \int_{-\infty}^{\infty} e^{iwt} \hat{\theta}(w) dw \tag{2.4}$$

$\mathcal{F}$ can be seen to be one to one and onto as well.

We denote by S_r the space containing S consisting of all C^r functions such that (2.1) is satisfied for $k \leq r$ but for all p. The topology is the same but with this restriction on k. Its Fourier transform will be denoted $\hat{S}_r$ and will consist of all C^∞ functions such that (2.1) is satisfied for all $p \leq r$ and all k.

A *tempered distribution* is an element of the dual space S' of S. This space is composed of all continuous linear functionals on S; functions from S to C (denoted by (T, θ), the value of T on θ), which are linear

$$(T, \alpha_1\theta_1 + \alpha_2\theta_2) = \alpha_2(T, \theta_1) + \alpha_2(T, \theta_2)$$

and continuous

$$\theta_n \to 0 \Rightarrow (T, \theta_n) \to 0$$

These are the ideal elements mentioned at the start of the chapter.

The limit of any bounded sequence of L^1 functions convergent in S' (i.e., convergent for each $\theta \in S$) defines such an element. Indeed, let $\{f_n\}$ be a sequence in $L^1(\mathbb{R})$ such that

$$\int_{-\infty}^{\infty} |f_n| \leq C \quad \text{and} \quad \{(f_n, \theta)\}$$

is a Cauchy sequence of complex numbers for each $\theta \in S$. Then we may define T as

$$(T, \theta) = \lim_{n \to \infty} (f_n, \theta).$$

This will be a linear functional and will be continuous as well since if $\theta_m \to \theta$,

$$|(f_n, \theta_m - \theta)| = \left| \int f_n(\theta_m - \theta) \right|$$

$$\leq \int |f_n| \sup_x |\theta_m(x) - \theta(x)|$$

$$\leq C \sup_x |\theta_m(x) - \theta(x)|$$

which converges to 0. Hence we can interchange the limits in

$$\lim_{m \to \infty} (T, \theta_m) = \lim_{m \to \infty} \lim_{n \to \infty} (f_n, \theta_m) = \lim_{n \to \infty} (f_n, \theta) = (T, \theta).$$

We could go further than this to prove that each Cauchy sequence converges in $\mathcal{S}'$ so that it is complete. We shall not do so (see, however [S, II, p.94].

Each locally integrable function f of polynomial growth belongs to $\mathcal{S}'$. The functional corresponding to it is

$$(f, \theta) = \int_{-\infty}^{\infty} f(t)\theta(t)dt$$

where we have used the same symbol f for the functional in $\mathcal{S}'$ and the function $f(t)$. This functional is clearly linear, and, since

$$f(t)/((t)^2 + 1)^r \in L^1$$

for some integer r, we see that

$$(f, \theta_n) = \int_{-\infty}^{\infty} \frac{f(t)}{(t^2 + 1)^r}(t^2 + 1)^r \theta_n(t)dt.$$

This is continuous since if $\theta_n \to 0$, $(t^2 + 1)^r \theta_n(t) \to 0$ uniformly and hence $(f, \theta_n) \to 0$.

Each $T \in \mathcal{S}'$ has a derivative defined by

$$(DT, \theta) = -(T, \theta'), \qquad \theta \in \mathcal{S}.$$

Thus even if T corresponds to a nowhere differentiable function, its derivative exists as a tempered distribution. By combining this with the previous example we see that each derivative of any order of a locally integrable function of polynomial growth belongs to $\mathcal{S}'$. In fact this characterizes $\mathcal{S}'$.

PROPOSITION 2.1
Let $T \in \mathcal{S}'$; then there exists a locally integrable function of polynomial growth $F(x)$ and an integer p such that

$$T = D^p F$$

(see [Ze, p.111]).

Thus, we may interpret elements of $\mathcal{S}'$ as generalizations of functions (in fact, they constitute examples of "generalized functions"), and shall adopt the same notation as for functions. The symbol (f, θ) will stand for either the value of the functional or, in the case of functions, the integral.

We already have the notation $\langle f, \overline{\theta} \rangle$ for the latter and hence $(f, \overline{\theta}) = \langle f, \theta \rangle$.

In addition to differentiation, we have a number of other operations defined on $\mathcal{S}'$. Each is defined by the same device. Any operation in $\mathcal{S}$ is translated to a corresponding operation in $\mathcal{S}'$ by first observing what happens to ordinary functions and then extending this to all of $\mathcal{S}'$. For example, translation T_α is defined on $\mathcal{S}'$ as

$$(T_\alpha f, \theta) = (f, T_{-\alpha}\theta)$$

since for ordinary functions

$$(T_\alpha f, \theta) = \int f(t-\alpha)\theta(t)dt = \int f(t)\theta(t+\alpha)dt = (f, T_{-\alpha}\theta).$$

Similarly for dilation we have

$$(D_\alpha f, \theta) = \frac{1}{\alpha}(f, D_{\frac{1}{\alpha}}\theta)$$

and for multiplication by a C^∞ function of polynomial growth $F(t)$,

$$(Ff, \theta) = (f, F\theta).$$

Of course we need to check for each definition that the result is in fact in $\mathcal{S}'$. But that again is clear in each of the cases.

The "delta function", δ_α, is defined simply as

$$(\delta_\alpha, \theta) = \theta(\alpha),$$

i.e., the unit point mass at α. It can also be given as the derivative of the Heaviside function

$$\delta_\alpha = DH_\alpha$$

where

$$H_\alpha(t) = \begin{cases} 1, & t > \alpha \\ 0, & t \leq \alpha \end{cases}.$$

Indeed, if $\theta \in \mathcal{S}$,

$$(H_\alpha, \theta') = \int_\alpha^\infty \theta'(t)dt = -\theta(\alpha) = -(\delta_\alpha, \theta).$$

Still a third way of defining δ_α is as the limit of a sequence of functions

$$\delta = \delta_0 = \lim_{n\to\infty} n\chi_{[0,n^{-1})}$$

where $\chi_{[0,n^{-1})}(t)$ is the characteristic function of $[0, n^{-1})$. Then $\delta_\alpha = T_\alpha\delta$. It is easy to check that this is consistent with the other definitions.

Other elements of $\mathcal{S}'$ include the pseudofunction such as $pv(\frac{1}{t})$ defined by

$$\left(pv\frac{1}{t},\theta\right) = cpv \int_{-\infty}^{\infty} \frac{\theta(t)}{t}\,dt$$

where cpv stands for Cauchy principal value.

In addition to having a derivative, each $f \in \mathcal{S}'$ has an antiderivative $f^{(-1)} \in \mathcal{S}'$. Let $\theta \in \mathcal{S}$ and define $\theta_1(t) = \theta(t) - \left(\int_{-\infty}^{\infty}\theta\right) h_0^2(t)$ where h_0 is the Hermite function of order 0. Then $\int_{-\infty}^{\infty}\theta_1 = 0$ and hence $\int_{-\infty}^{t}\theta_1 \in \mathcal{S}$. We now define $f^{(-1)}$ by

$$(f^{(-1)},\theta) := -(f, \int_{-\infty}^{t}\theta_1) + C(1,\theta)$$

where C is arbitrary. Then

$$(Df^{(-1)},\theta) = -(f^{(-1)},\theta') = (f, \int_{-\infty}^{t}\theta_1(\theta)) + C(1,\theta')$$

$$= (f, \int_{-\infty}^{t}\theta') = (f,\theta)$$

since $(1,\theta') = 0$ and the θ_1 corresponding to θ' equals θ'.

The <u>support</u> of a function $\theta \in \mathcal{S}$ is the smallest closed set outside of which θ is identically zero. Two distributions f and $g \in \mathcal{S}'$ are said to be equal in an open set Ω if their difference is zero on each θ with support in Ω, i.e., if

$$(f,\theta) = (g,\theta), \quad \text{supp } \theta \subset \Omega.$$

In particular f could be zero on some such Ω. The <u>support</u> of f then is the smallest closed set K such that $f = 0$ on the complement of K. Clearly δ has support on the single point $\{0\}$, as do all of its derivatives. In fact, the only $f \in \mathcal{S}'$ with point support are linear combinations of δ and its derivatives ([Ze, p.81]).

The delta function also satisfies the useful <u>sifting property</u>. For $F(t)$ a continuous function at $t = \alpha$, we have

$$F(t)\delta_\alpha(t) = F(\alpha)\delta_\alpha(t).$$

This is clear for $F \in C^\infty$ but may be shown for other functions by using the sequence approach to δ.

There is also an intimate connection between δ and convolution. For a tempered distribution g with compact support, the convolution with any $f \in \mathcal{S}'$ can be defined. (The convolution of two $L'(\mathbb{R})$ functions is defined to be $f * g(t) := \int_{-\infty}^{\infty} f(t - s)g(s)ds$.)

We first take the convolution of g with $\theta \in \mathcal{S}$ defined as

$$(g * \theta)(t) = (g, T_t \tilde{\theta})$$

where $\tilde{\theta}(t) = \theta(-t)$. Then $g * \theta$ is a continuously differentiable function which belongs to $\mathcal{S}$. The convolution of f with g is then defined as

$$(f * g, \theta) := (f, \widetilde{g * \tilde{\theta}}).$$

There is a lot of checking to do here but it all works if g has compact support.

Since δ has compact support, its convolution exists for every $f \in \mathcal{S}'$ and in fact

$$f * \delta = f.$$

These distributions of compact support have a simple characterization. If g has support on a compact interval K such that $0 \in K$, then there is a continuous function G with support on K and an integer p such that

$$g = D^p G + \sum_{i=0}^{p-1} c_i \delta^{(i)}.$$

This is proved by observing that $g = D^p F$ for some continuous function F of polynomial growth. But $D^p F$ is zero on the complement of K, and therefore F is a polynomial of degree $\leq p - 1$ on this complement. Take $G(t) = F(t)-$ this polynomial for t negative and similarly for t positive. Then $D^p G - g$ has support on $\{0\}$ and must be a linear combination of δ and its derivatives.

2.2 Fourier transforms

The definition of Fourier transform may be extended from $\mathcal{S}$ to $\mathcal{S}'$ by the same device as the other operations. For $f \in \mathcal{S}'$ we define $\hat{f} = \mathcal{F}f$ to be that element of $\mathcal{S}'$ given by

$$(\hat{f}, \theta) = (f, \hat{\theta}).$$

It is easy to check that this is a linear functional and that it is continuous. Moreover the map $f \mapsto \hat{f}$ is a continuous linear transformation from $\mathcal{S}'$ onto $\mathcal{S}'$ whose inverse exists and is continuous as well.

Some examples of Fourier transforms on $\mathcal{S}'$ are:

$$\mathcal{F}\delta_\alpha = e^{-iw\alpha},$$

$$\mathcal{F}\sum_k \delta(t - 2\pi k) = \sum_n \delta(w - n),$$

$$\mathcal{F}(t^k e^{-it\tau}) = i^k 2\pi \delta^{(k)}(w + \tau),$$

$$\mathcal{F}\left(pv\frac{1}{t}\right) = -\pi i \, \mathrm{sgn} w,$$

$$\mathcal{F}\left(\sum_n a_n e^{int}\right) = 2\pi \sum_n a_n \delta(w - n),$$

$$\mathcal{F}\left(\sum_n a_n e^{int}\chi_{[-\pi,\pi]}(t)\right) = 2\pi \sum_n a_n \frac{\sin \pi(w - n)}{\pi(w - n)},$$

where $\chi_{[-\pi,\pi]}$ is the characteristic function of $[-\pi, \pi]$, $\{a_n\} \in \ell^2$. Each of these may be found by using the definition. For example

$$(\mathcal{F}\delta_\alpha, \theta) = (\delta_\alpha, \hat\theta) = \hat\theta(\alpha) = \int_{-\infty}^{\infty} e^{-iw\alpha}\theta(w)dw = (e^{-iw\alpha}, \theta).$$

Furthermore, this definition of Fourier transform is consistent with the usual definition for functions in $L^1(\mathbb{R})$. Since elements of $\mathcal{S}'$ may be approximated by such functions, the usual definition may be used for $f \in \mathcal{S}'$ provided it makes sense. Again $\mathcal{F}\delta_\alpha$ may be found by using

$$\mathcal{F}\delta_\alpha = (\delta_\alpha, e^{-iw\cdot}) = e^{-iw\alpha}.$$

Still another approach involves using Hermite functions $h_n(x)$, $n = 0, 1, \cdots$. It is well known that $\hat h_0(w) = \sqrt{2\pi}h_0(w)$, i.e., $h_0(w) = \frac{1}{\sqrt{\pi}}e^{-w^2/2}$ is an eigenfunction of the Fourier transform. Since $\mathcal{F}(xf)(w) = i\hat f'(w)$ and $\mathcal{F}(f')(w) = iw\hat f(w)$, the equation

$$(x - D)h_n = \sqrt{2n + 2}h_{n+1}$$

can be used to find the Fourier transform of all h_n. For example, $h_1 = \frac{1}{\sqrt{2}}(x - D)h_0$ and therefore

$$\hat h_1(w) = \frac{1}{\sqrt{2}}i(\hat h_0'(w) - w\hat h_0(w))$$

$$= -\frac{i}{\sqrt{2}}\left(w - \frac{d}{dw}\right)\sqrt{2\pi}h_0(w)$$

$$= -\sqrt{2\pi}i\, h_1(w).$$

Similarly for all values of n we find

$$\hat{h}_n(w) = \sqrt{2\pi}(-i)^n h_n(w),$$

i.e., all h_n are eigenfunctions of the Fourier transform. Thus if f has an expansion

$$f = \Sigma a_n h_n$$

then

$$\hat{f} = \sqrt{2\pi}\Sigma a_n(-i)^n h_n. \tag{2.5}$$

It's fortunate that all $f \in \mathcal{S}'$ have such an expansion (see Chapter 6).

THEOREM 2.2
Let $f \in \mathcal{S}'$; then there is a $p \in \mathbb{N}$ such that

$$a_n = (f, h_n) = O(n^p),$$

and

$$f = \sum_n a_n h_n$$

with convergence in the sense of $\mathcal{S}'$.

Again we may find $\mathcal{F}\delta_\alpha$ by this device.

$$\delta_\alpha = \sum_n (\delta_\alpha, h_n)h_n = \sum_n h_n(\alpha)h_n$$

and

$$(\mathcal{F}\delta_\alpha)(w) = \sqrt{2\pi}\sum_n (-i)^n h_n(\alpha)h_n(w)$$

$$= \sum_n (e^{-i\cdot\alpha}, h_n)h_n(w) = e^{-iw\alpha}.$$

2.3 Periodic distributions

Some tempered distributions are periodic of period 2π, i.e., $T_{2\pi}f = f$ where $T_{2\pi}$ is the operator of translation by an amount 2π. Such tempered distributions have a Fourier series with coefficient c_n given by

$$c_n = \frac{1}{2\pi}\left[T_{2n}\left(fe^{int}\right)^{(-1)} - \left(fe^{int}\right)^{(-1)}\right]. \tag{2.6}$$

While this seems a rather awkward definition, it reduces to

$$c_n = \frac{1}{2\pi} \int_t^{t+2\pi} f(x)e^{-inx}\,dx$$

when f is an ordinary function. In fact, this may be taken as the definition for $f \in \mathcal{S}'$ provided it is interpreted as (2.6).

The resulting c_n is a constant since

$$\frac{dc_n}{dt} = \frac{1}{2\pi}[T_{2\pi}(fe^{-int}) - (fe^{-int})] = 0$$

by the periodicity of fe^{-int}. We then associate with f the Fourier series

$$\sum_n c_n e^{int}.$$

THEOREM 2.3
Let $f \in \mathcal{S}'$ be periodic of period 2π; let c_n be given by (2.6); then there is a $p \in \mathbb{N}$ such that
$$c_n = O(|n|^p)$$

and
$$f = \sum c_n e^{int}$$

with convergence in the sense of $\mathcal{S}'$.

(See Chapter 4).

2.4   Analytic representations

The elements of $\mathcal{S}'$ are not necessarily ordinary functions but can be represented by pairs of ordinary functions in the complex plane. This is the analytic representation given by

$$f^{\pm}(z) = \frac{1}{2\pi i} \int_{-\infty}^{\infty} \frac{f(x)}{x - z}\,dx, \qquad \text{Im } z \neq 0$$

for $f \in L^2(\mathbb{R})$. The functions $f^{\pm}(z)$ are analytic in the upper half-plane $(+)$ and in the lower half-plane $(-)$ and

$$\lim_{\epsilon \to 0}(f^+(x + i\epsilon) - f^-(x - i\epsilon)) = f(x) \qquad \text{a.e.}$$

The definition may be first extended to elements of $\mathcal{S}'$ whose Fourier transforms are continuous functions of polynomial growth by

$$f^+(z) = \frac{1}{2\pi} \int_0^\infty e^{iwz} \hat{f}(w)\,dw \qquad \mathrm{Im}\, z > 0,$$

$$f^-(z) = \frac{1}{2\pi} \int_{-\infty}^0 e^{iwz} \hat{f}(w)\,dw \qquad \mathrm{Im}\, z < 0. \tag{2.7}$$

The integrals exist as ordinary Lebesgue integrals since the integrand in the first, for example, is dominated by

$$e^{-w\,\mathrm{Im}\,z}\,\hat{f}(w).$$

Furthermore, the complex derivative of the integrand is also integrable and hence $f^+(z)$ is analytic in the upper and $f^-(z)$ in the lower half-plane. By the inversion formula for Fourier transforms we get

$$(f^+(x+i\epsilon) - f^-(x-i\epsilon),\ \theta(x)) \to (f, \theta) \qquad \text{as } \epsilon \to 0 \tag{2.8}$$

for all $\theta \in \mathcal{S}$, [Br, p.87]. Since all $f \in \mathcal{S}'$ are finite order derivatives of continuous functions of polynomial growth [Ze], we may represent $\hat{f}$ by

$$\hat{f}(w) = D^p \hat{g}(w).$$

This defines a pair of analytic functions

$$f^\pm(z) = (-iz)^p g^\pm(z),$$

which also satisfy (2.8) [Br, p.90].

Some examples of analytic representations are

$$\delta^\pm(z) = -\frac{1}{2\pi i z}, \qquad \mathrm{Im}\, z \neq 0,$$

$$\left(pv\frac{1}{t}\right)^+(z) = \frac{1}{2z}, \qquad \mathrm{Im}\, z > 0,$$

$$\left(pv\frac{1}{t}\right)^-(z) = \frac{-1}{2z}, \qquad \mathrm{Im}\, z < 0,$$

$$\begin{aligned} p^+(z) &= p(z)/2, \quad \mathrm{Im}\, z > 0, \\ p^-(z) &= -p(z)/2, \quad \mathrm{Im}\, z < 0, \end{aligned} \qquad p(t) \text{ a polynomial,}$$

$$h^\pm(z) = -\frac{1}{2\pi i}\log(-z), \qquad \mathrm{Im}\, z \neq 0, \qquad h(t) = \begin{cases} 1 & t > 0 \\ 0 & t \leq 0. \end{cases}$$

2.5   Sobolev spaces

A convenient Hilbert space structure to use on certain subsets of S' is that of *Sobolev spaces*. The Sobolev space H^α, $\alpha \in \mathbb{R}$ consists of all $f \in S'$ such that

$$\int_{-\infty}^{\infty} |\hat{f}(w)|^2 (w^2 + 1)^\alpha dw < \infty.$$

For $\alpha = 0$, this is just $L^2(\mathbb{R})$; for $\alpha = 1, 2, \ldots H^\alpha$ is composed of ordinary $L^2(\mathbb{R})$ functions that are $(\alpha - 1)$ times differentiable and whose α-th derivative is in $L^2(\mathbb{R})$; for $\alpha = -1, -2, \ldots$, H^α contains all distributions with point support of order $\leq \alpha$ (i.e., $\delta, \delta', \ldots, \delta^{(\alpha-1)}$). Clearly $H^\alpha \supset H^\beta$ when $\alpha < \beta$.

The inner product of f and $g \in H^\alpha$ is defined as

$$\langle f, g \rangle_\alpha = \frac{1}{2\pi} \int_{-\infty}^{\infty} \hat{f}(w)\overline{\hat{g}(w)}(w^2 + 1)^\alpha dw.$$

It is complete with respect to this inner product and hence is a Hilbert space.

Concluding remark

This has been a somewhat abbreviated introduction to tempered distributions. A more detailed but still simple introduction is given by Zemanian [Ze]. A much more complete treatment is found in the five volumes of Gelfand, Shilov [G-S].

These tempered distributions constitute a subspace of the space of distributions $\mathcal{D}'$ of L. Schwartz [S], which we have not discussed. However we shall often refer to particular elements of S' as "distributions," which is proper because of this containment relation.

2.6   Problems

1. Check that the definition of convolution of $f \in S'$ with $g \in S'$ with compact support defines an element of S'.

2. Show that $f(x)\delta(x-a) = f(a)\delta(x-a)$ for a continuous function f of polynomial growth.

3. Establish the formula

$$D^p[f(x)\delta(x)] = \sum_{k=0}^{p} \binom{p}{k} f^{(k)}(x)\delta^{(p-k)}(x)$$

and use it to find

$$f(x)\delta^{(p)}(x)$$

for $f \in C^p$ and of polynomial growth.

4. Use the result in Problem 3 to show that

$$x^n\delta^{(m)}(x) = (-1)^n \frac{m!}{(m-n)!}\delta^{(m-n)}(x), \quad m \geq n.$$

5. Show that

$$\delta^{(k)}(2^m x) = 2^{-m(k+1)}\delta^{(k)}(x).$$

6. Show that if $\theta_n \to 0$ in S so does its Fourier transform $\hat{\theta}_n$.

7. Show that the examples in Section 2.2 have the Fourier transforms indicated. Sometimes it will help to use (2.4) to simplify the calculations.

8. Find the Fourier transform of the periodic distribution

$$\delta^*(x) = \sum_{k=-\infty}^{\infty} \delta(x - 2\pi k).$$

9. Find the Fourier series of $\delta^*(x)$ given in Problem 8.

10. Show that the analytic representation $\delta^{\pm}(z)$ of δ satisfies

$$\lim_{\epsilon \to 0} \delta^+(x + i\epsilon) - \delta^-(x - i\epsilon) = \delta(x).$$

11. Show that the Sobolev space H^{-1} contains $\delta(x)$ but not $\delta'(x)$.

3

An Introduction to Orthogonal Wavelet Theory

The structure of systems of orthogonal wavelets is similar to that of the prototypes in Chapter One. However, both they and their Fourier transforms have a certain degree of smoothness that the prototypes do not have. It was the discovery of smooth orthogonal wavelets by Meyer [M], Daubechies [D1], and others that led to the explosion of activity in the subject.

Wavelets began not as an orthogonal system but, rather, as an integral transform involving dilations and translations of a fixed function. These were found to give a better representation for certain seismic signals than Fourier methods [Mo]. The original theory was soon put into a rigorous framework by a number of mathematicians [G-M], [M]. Next, a discrete nonorthogonal representation based on frames evolved [D1]. Finally, the orthogonal theory was discovered [D2], [M1]. Many of the techniques, based on harmonic analysis, had been around for a long time but had not been directed toward the particular problem associated with wavelets.

The wavelet basis is composed of functions $\psi_{mn}(t)$ given by

$$\psi_{mn}(t) = 2^{m/2}\psi(2^m t - n), \qquad m, n \in \mathbb{Z},$$

where ψ is a fixed function in $L^2(\mathbb{R})$, the so-called "mother wavelet". It's not clear that such an orthonormal basis consisting of dilations and translations of a smooth function exists. Much of the theory has been devoted to constructing $\psi(t)$ for which such a basis exists. This construction is usually based on another function $\phi(t)$, the "scaling function".

3.1 Multiresolution analysis

We begin with a scaling function ϕ, a real valued function on $\mathbb{R}$ which is r times differentiable and whose derivatives are continuous and rapidly decreasing. That is, ϕ satisfies

$$|\phi^{(k)}(t)| \le C_{pk}(1+|t|)^{-p}, \quad k = 0, 1, \ldots, r; \quad p \in \mathbb{Z}, \quad t \in \mathbb{R} \quad (3.1)$$

and hence $\phi \in \mathcal{S}_r$. In order for it to qualify as a scaling function, there must be associated with ϕ a multiresolution analysis of $L^2(\mathbb{R})$, i.e., a nested sequence of closed subspaces $\{V_m\}_{m \in \mathbb{Z}}$ such that

$$
\begin{array}{lll}
\text{(i)} & \{\phi(t-n)\} \text{ is an orthonormal basis of } V_0 & \\
\text{(ii)} & \cdots \subset V_{-1} \subset V_0 \subset V_1 \subset \cdots \subset L^2(\mathbb{R}) & \\
\text{(iii)} & f \in V_m \Leftrightarrow f(2\cdot) \in V_{m+1} & (3.2) \\
\text{(iv)} & \bigcap_m V_m = \{0\}, \quad \overline{\bigcup_m V_m} = L^2(\mathbb{R}). &
\end{array}
$$

Clearly $\{\sqrt{2}\phi(2t-n)\}$ is an orthonormal basis for V_1 since the map $f \vdash \sqrt{2}f(2\cdot)$ is an isometry from V_0 onto V_1. Since $\phi \in V_1$, it must have an expansion

$$\phi(t) = \sum_k c_k \sqrt{2}\phi(2t-k), \quad \{c_k\} \in \ell^2, \quad t \in \mathbb{R}. \quad (3.3)$$

Example 1

A simple example, which however does not satisfy (3.1), is given by $\phi(t) = \chi_{[0,1]}(t)$, the characteristic function of the unit interval. Its translates $\{\phi(t-n)\}$ are orthonormal and V_0 is composed of piecewise constant functions. This is prototype I of Chapter 1.

It is not obvious that a multiresolution analysis exists for ϕ other than this example and prototype II. In order to find other ϕ's, three approaches have been used. One begins with an existing multiresolution analysis of subspaces $\{V_m\}$ defined in some other way and then tries to find an orthogonal basis. In particular, V_0 may be defined by means of a Riesz basis of translates of a fixed function $\{\theta(t-n)\}$. Then an orthogonalization procedure due to Lemarié and Meyer [L-M] is used to find an orthogonal $\{\phi(t-n)\}$. This is not the familiar Gram-Schmidt orthogonalization, which, while it would give an orthogonal basis, would not guarantee that it be composed of translates of a fixed function.

The second method, due to Daubechies [D2], consists of choosing the dilation coefficients $\{c_k\}$ in (3.3) such that all the required conditions are satisfied. The third approach is similar; it consists of choosing the Fourier transform of $\phi(t)$ in such a way that it has compact support, and the transformed versions of (3.2) and (3.3) are satisfied.

Example 2

The simplest example of the orthogonalization approach leads to the *Franklin wavelets*. It starts with the "hat" function $\theta(t) = (1 - |t - 1|)\chi_{[0,2]}(t)$. This corresponds to a V_0 composed of continuous piecewise linear functions with breaks at the integers. Its Fourier transform is

$$\hat{\theta}(w) = \int_{-\infty}^{\infty} e^{-iwt}\theta(t)dt = \left[\frac{1 - e^{-iw}}{iw}\right]^2 = \frac{e^{-iw}\sin^2 w/2}{(w/2)^2}. \qquad (3.4)$$

To find the orthogonal ϕ we use the isometry property of the Fourier transform:

$$\delta_{on} = \int_{-\infty}^{\infty} \phi(t - n)\phi(t)dt = \frac{1}{2\pi}\int_{-\infty}^{\infty} \hat{\phi}(w)e^{-iwn}\overline{\hat{\phi}(w)}dw$$

$$= \frac{1}{2\pi}\sum_{k=-\infty}^{\infty}\int_{0}^{2\pi}\left|\hat{\phi}(w + 2\pi k)\right|^2 e^{-iwn}dw$$

$$= \frac{1}{2\pi}\int_{0}^{2\pi}\sum_{k}\left|\hat{\phi}(w + 2\pi k)\right|^2 e^{-iwn}dw.$$

This is the integral that gives the Fourier coefficient of the periodic function $\left|\hat{\phi}_2^*(w)\right|^2 := \sum_{k}\left|\hat{\phi}(w + 2\pi k)\right|^2$, which by orthogonality must be δ_{on}. Hence $\left|\hat{\phi}^*(w)\right|^2$, which is equal to its Fourier series, is identically equal to 1, i.e.,

$$\sum |\hat{\phi}(w + 2\pi k)|^2 = 1. \qquad (3.5)$$

Since $\phi \in V_0$, we have $\phi(t) = \sum a_n \theta(t - n)$ for some $\{a_n\} \in \ell^2$. Again by taking Fourier transforms we find

$$\hat{\phi}(w) = \sum_{n} a_n e^{-iwn}\hat{\theta}(w) = \alpha(w)\hat{\theta}(w) \qquad (3.6)$$

and hence

$$\left|\hat{\phi}_2^*(w)\right|^2 = |\alpha(w)|^2\left|\hat{\theta}^*(w)\right|^2 = 1. \qquad (3.7)$$

$\left|\hat{\theta}_2^*(w)\right|^2$ can be found in closed form by summing the series obtained from (3.4)

$$\left|\hat{\theta}_2^*(w)\right|^2 = \sum_{k}\left|\frac{\sin(w + 2\pi k)/2}{(w + 2\pi k)/2}\right|^4 = (\sin w/2)^4 \sum_{k}\frac{1}{\left(\frac{w}{2} + \pi k\right)^4}.$$

This may be done by using the Fourier series of e^{-ixt}

$$e^{-ixt} = \sum_k \frac{\sin \pi(x+k)}{\pi(x+k)} e^{ikt}, \quad -\pi < t < \pi.$$

Then by Parseval's equality we have

$$1 = \sum_k \frac{\sin^2 \pi x}{\pi^2 (x+k)^2}$$

or

$$\frac{1}{\sin^2 \pi x} = \sum_k \frac{1}{(\pi x + \pi k)^2}.$$

By differentiating both sides twice we obtain

$$\frac{2\pi^2 (\sin^2 \pi x + 3\cos^2 \pi x)}{\sin^4 \pi x} = \sum_k \frac{6\pi^2}{(\pi x + \pi k)^4}.$$

We then replace πx by $\frac{w}{2}$ to find

$$\sum_k \frac{1}{\left(\frac{w}{2} + \pi k\right)^4} = \frac{1 - \frac{2}{3}\sin^2 \frac{w}{2}}{\sin^4 \frac{w}{2}}.$$

Hence by taking the positive square root and multiplying by the periodic factor e^{-iw} we find

$$\hat{\theta}_2^*(w) = e^{-iw} \left\{ 1 - \frac{2}{3}\sin^2 \frac{w}{2} \right\}^{\frac{1}{2}}.$$

By substituting the result in (3.7), solving for $\alpha(w)$, and substituting it in (3.6), we find ϕ to be

$$\hat{\phi}(w) = \frac{\sin^2 w/2}{(w/2)^2} \left\{ 1 - \frac{2}{3}\sin^2 w/2 \right\}^{-\frac{1}{2}}. \tag{3.8}$$

Since $\left\{ 1 - \frac{2}{3}\sin^2 w/2 \right\}^{-1/2}$ is an analytic function in a strip about the real axis and is of period 2π, its Fourier series has coefficients that are exponentially decreasing and

$$\hat{\phi}(w) = e^{iw} \left[\frac{1 - e^{-iw}}{iw} \right]^2 \sum_k a_k e^{iwk}.$$

The inverse Fourier transform is

$$\phi(t) = \sum_k a_k \theta(t + k + 1), \qquad a_k = O(e^{-a|k|}).$$

The coefficients a_k may be found explicitly by expanding

$$\left(1 - \frac{2}{3}\sin^2 w/2\right)^{-\frac{1}{2}} = \sqrt{\frac{3}{2}}\left(1 + \frac{\cos w}{2}\right)^{-\frac{1}{2}}$$

in powers of $\frac{\cos w}{2}$.

Example 3

The *Daubechies wavelets* are based on the other approach. We look for $\{c_k\}$ in (3.3) such that the orthogonality condition is satisfied. This becomes

$$\int \phi(t)\phi(t - n)dt = \sum_k c_k \int \sqrt{2}\phi(2t - k)\sum_j c_j \sqrt{2}\phi(2t - 2n - j)dt$$

$$= \sum c_k c_{k-2n} = \delta_{on}, \qquad n \in \mathbb{Z}.$$

$$(3.9)$$

The conditions $\hat{\phi}(o) = 1$, $\hat{\phi}(2\pi) = 0$ implied by orthogonality (Section 3.4) become

$$\sum_k \frac{c_k}{\sqrt{2}} = 1, \qquad \sum \frac{c_k}{\sqrt{2}}(-1)^k = 0 \qquad (3.10)$$

if $\hat{\phi}(\pi) \neq 0$. These are redundant since both are based on orthogonality. However, there are three independent equations given by (3.9) for $n = 0$ and (3.10). If c_k is chosen by

$$c_0 = \nu(\nu - 1)/(\nu^2 + 1)\sqrt{2},$$
$$c_1 = (1 - \nu)/(\nu^2 + 1)\sqrt{2},$$
$$c_2 = (\nu + 1)/(\nu^2 + 1)\sqrt{2},$$
$$c_3 = \nu(\nu + 1)/(\nu^2 + 1)\sqrt{2},$$

where $\nu \in \mathbb{R}$ (and all the other $c_k = 0$), then (3.9) and (3.10) will be satisfied.

If we require some degree of smoothness we can add the moment condition

$$\int_{-\infty}^{\infty} t\,\psi(t)dt = 0,$$

which can be shown (see Section 3.4) to lead to the equation

$$\sum_k c_k(-1)^k k = 0.$$

There are then two solutions to these four equations given by $\nu = \pm 1/\sqrt{3}$.

This does not directly give $\phi(t)$, which rather must be found recursively by starting with, say, $\phi_0(t) = \chi_{[0,1)}(t)$ and then defining

$$\phi_n(t) = \sum_k c_k \sqrt{2} \phi_{n-1}(2t - k). \tag{3.11}$$

This sequence converges to a continuous function when $\nu = \pm 1/\sqrt{3}$ but in general, does not, although it does converge weakly [V], whatever the choice of $\{c_n\}$.

The dilation equation (3.3) may also be expressed in terms of Fourier transforms. It is

$$\hat{\phi}(w) = \sum_k \frac{c_k}{\sqrt{2}} \hat{\phi}\left(\frac{w}{2}\right) e^{-iwk/2}.$$

and if we denote by

$$m_o\left(\frac{w}{2}\right) = \sum_{m=0}^{\infty} \frac{c_k}{\sqrt{2}} e^{-iwk/2},$$

the orthogonality condition (3.5) becomes

$$\begin{aligned}
\left|\hat{\phi}_2^*(w)\right| &= \sum_k \left|m_o\left(\frac{w+2\pi k}{2}\right)\right|^2 \left|\hat{\phi}\left(\frac{w+2\pi k}{2}\right)\right|^2 \\
&= \sum \left|m_o\left(\frac{w}{2}+2\pi j\right)\right|^2 \left|\hat{\phi}\left(\frac{w}{2}+2\pi j\right)\right|^2 \\
&\quad + \left|m_o\left(\frac{w}{2}+(2j+1)\pi\right)\right|^2 \left|\hat{\phi}\left(\frac{w}{w}+(2j+1)\pi\right)\right|^2 \\
&= \left|m_o\left(\frac{w}{2}\right)\right|^2 \left|\hat{\phi}^*\left(\frac{w}{2}\right)\right|^2 + \left|m_o\left(\frac{w}{2}+\pi\right)\right|^2 \left|\hat{\phi}^*\left(\frac{w}{2}+\pi\right)\right|^2
\end{aligned}$$

or

$$\left|m_o\left(\frac{w}{2}\right)\right|^2 + \left(m_o\left(\frac{w}{2}+\pi\right)\right|^2 = 1.$$

This, then, is a necessary condition for orthogonality, which we use later to construct more examples.

3.2 Mother wavelet

Once we have the scaling function $\phi(t)$, we may use it to construct the "mother wavelet" $\psi(t)$. It must be chosen such that $\{\psi(t-n)\}$ is an orthonormal basis of the space W_0, given by the orthogonal complement of V_0 in V_1. Then

$$V_1 = V_0 \oplus W_0$$

If such a $\psi(t)$ can be found, then $2^{m/2}\psi\left(2^{m/2}t - n\right) = \psi_{nm}(t)$ is an orthonormal basis of W_m, the dilation space of W_0. Furthermore W_m will be the orthogonal complement of V_m in V_{m+1}. Hence we have

$$V_{m+1} = V_m \oplus W_m = \cdots = V_0 \oplus W_0 \oplus W_1 \oplus \cdots \oplus W_m,$$

and since $\bigcup_m V_m$ is dense in $L^2(\mathbb{R})$, we may take the limit as $m \to \infty$ to obtain

$$V_0 \oplus \left(\bigoplus_{m=0}^{\infty} W_m \right) = L^2(\mathbb{R}).$$

We may also go in the other direction to write

$$V_0 = V_{-1} \oplus W_{-1} = V_{-k} \oplus W_{-k} \oplus \cdots \oplus W_{-1}.$$

We again take the limit as $k \to \infty$. Since $\bigcap_m V_m = \{o\}$, it follows that $V_{-k} \to \{o\}$ and it follows that

$$\bigoplus_{m \in Z} W_m = L^2(\mathbb{R}),$$

and hence $\{\psi_{nm}\}_{n,m \in Z}$ is an orthonormal basis of $L^2(\mathbb{R})$.

Again there are two methods of finding $\psi(t)$. One uses the orthogonalization procedure similar to that illustrated in Example 2. But $\psi(t)$ must also be orthogonal to $\phi(t - n)$. Hence the two conditions

$$\text{(a)} \quad \sum_k \hat{\psi}(w + \pi k)\overline{\hat{\phi}(w + 2\pi k)} = 0$$

$$\text{(b)} \quad \sum_k \left| \hat{\psi}(w + 2\pi k) \right|^2 = 1$$

$$(3.12)$$

must be satisfied. Since $\psi(t) \in V_1$, it has an expansion similar to (3.3)

$$\psi(t) = \sum_k d_k \sqrt{2}\phi(2t - k) \tag{3.13}$$

and hence

$$\hat{\psi}(w) = m_1\left(\frac{w}{2}\right)\hat{\phi}\left(\frac{w}{2}\right). \tag{3.14}$$

This may be substituted into (3.12) and the two equations solved for $m_1\left(\frac{w}{2}\right)$, which is then substituted back in (3.14) to find $\hat{\psi}(w)$, i.e., we find

$$\text{(a)} \quad m_1\left(\frac{w}{2}\right)\overline{m_o\left(\frac{w}{2}\right)} + m_1\left(\frac{w}{2}+\pi\right)\overline{m_o\left(\frac{w}{2}+\pi\right)} = 0$$

$$\text{(b)} \quad \left|m_1\left(\frac{w}{2}\right)\right|^2 + \left|m_1\left(\frac{w}{2}+\pi\right)\right|^2 = 0 \tag{3.15}$$

These two equations may then be solved simultaneously for $\left|m_1\left(\frac{w}{2}\right)\right|^2$ and then a square root taken. However, the positive square root doesn't work, but

$$m_1\left(\frac{w}{2}\right) = e^{-iw/2}\overline{m_o\left(\frac{w}{2}+\pi\right)}$$

does. Of course, there are other solutions.

The other approach involving the dilation equation is straightforward; $\psi(t)$ is defined as

$$\psi(t) := \sqrt{2}\sum_k c_{1-k}(-1)^k\phi(2t-k). \tag{3.16}$$

Then it is merely a matter of checking that the two orthogonality conditions are satisfied [D2, p.944].

By taking the Fourier transform of (3.16) we obtain the expression

$$\hat{\psi}(w) = 1/\sqrt{2}\sum_k c_{1-k}e^{-i(\frac{w}{2}+\pi)k}\hat{\phi}(\frac{w}{2})$$

$$= 1/\sqrt{2}\,e^{-i\frac{w}{2}}\sum_j c_j e^{i(\frac{w}{2}+\pi)j}\hat{\phi}(\frac{w}{2})$$

$$= e^{-i\frac{w}{2}}\overline{m_0(\frac{w}{2}+\pi)}\hat{\phi}(\frac{w}{2}),$$

which agrees with (3.15).

Examples

In Example 1, $\psi(t) = \phi(2t) - \phi(2t-1) = \begin{cases} 1, & 0 \le t < 1/2 \\ -1, & 1/2 \le t < 1 \\ 0, & \text{otherwise} \end{cases}$.

In Example 2, we find from (3.8) that

$$m_0(\frac{w}{2}) = \frac{2 + 3\cos w/2 + \cos^2 w/2}{1 + 2\cos^2 w/2}$$

and hence that

$$\hat{\psi}(w) = e^{-i\frac{w}{2}}\frac{2 - 3\cos w/2 + \cos^2 w/2}{1 + 2\cos^2 w/2}\hat{\phi}(\frac{w}{2}).$$

In Example 3, the best we can do is use the coefficients c_0, c_1, c_2, and c_3 in Equation (3.16). Again as in $\hat{\phi}(w)$, there is no formula, but there is a recursion algorithm based on Equation (3.16).

Example 4

The *B-spline of order n* is the function $\theta_n(t)$ obtained by convolving the Haar scaling function with itself n-times. Its Fourier transform is

$$\hat{\theta}_n(w) = \left[\frac{1 - e^{-iw}}{iw}\right]^{n+1}.$$

For $n = 1$, these give the Franklin wavelets of Example 2. For other values of n, $\theta_n(t) \in C^n$, has support on $[0, n+1]$ and is a polynomial of degree $n+1$ in each interval $(k, k+1)$ where $k \in \mathbb{Z}$ [C], [M]. They may be orthogonalized by the same trick as in Example 2 to obtain

$$\hat{\phi}(w) = \left(\frac{\sin w/2}{w/2}\right)^{n+1} [P(\cos w/2)]^{-1/2}$$

where P is a polynomial of degree $2n$, strictly positive on $[0, 1]$, when n is odd. When n is even the expression becomes

$$\hat{\phi}(w) = \left(\frac{\sin w/2}{w/2}\right)^{n+1} e^{-i\frac{w}{2}} [P(\cos w/2)]^{-1/2}.$$

The mother wavelet is obtained from (3.16). These are the Battle-Lemarie wavelets.

Example 5

For this example the role of ϕ and $\hat{\phi}$ in Example 1 is reversed. The *Shannon* scaling function prototype II of Chapter 1 is

$$\phi(t) = \frac{\sin \pi t}{\pi t}$$

or

$$\hat{\phi}(w) = \chi_{[-\pi, \pi)}(w).$$

In this case ϕ is not rapidly decreasing, but ϕ has other nice properties. In particular for $f \in V_0$, which in this case is composed of π band limited functions, we have the *Shannon Sampling Theorem*

$$f(t) = \sum_{n=-\infty}^{\infty} f(n)\phi(t - n).$$

A possible mother wavelet is given by

$$\psi(t) = \frac{\sin \pi(t - \frac{1}{2}) - \sin 2\pi(t - \frac{1}{2})}{\pi(t - \frac{1}{2})};$$

another is given in Section 1.2.3.

Example 6
A smoothed version of Example 5 is the *Meyer wavelet*. Here $\phi(t) \in C^\infty$, $\hat{\phi}(w) = 1$ for $\frac{-2\pi}{3} \le w \le \frac{2\pi}{3}$, $\hat{\phi}(w) = 0$ for $|w| \ge \frac{4\pi}{3}$, and in between it satisfies the condition

$$[\hat{\phi}(w - \pi)]^2 + [\hat{\phi}(w + \pi)]^2 = 1.$$

(See [M, p.60]). In this case $\phi \in \mathcal{S}$.

Example 7
Examples 5 and 6 can both be generalized as follows:

Let P be a probability measure with support in $[-\epsilon, \epsilon] \subset [-\frac{\pi}{3}, \frac{\pi}{3}]$ and define $\phi(t)$ as the function whose Fourier transform is

$$\hat{\phi}(w) = \left[\int_{w-\pi}^{w+\pi} dP \right]^{\frac{1}{2}},$$

the nonnegative square root of the integral. Then $\hat{\phi}$ has support in $[-\frac{4\pi}{3}, \frac{4\pi}{3}]$, and $\hat{\phi}(w) = 1$ for $|w| < \frac{2\pi}{3}$. The orthogonality condition is satisfied since

$$\sum_k |\hat{\phi}(w + 2\pi k)|^2 = \sum_k \int_{w+(2k-1)\pi}^{w+(2k+1)\pi} dP = \int_{-\infty}^{\infty} dP = 1$$

as is the dilation condition

$$\hat{\phi}(w) = m_o\left(\frac{w}{2}\right) \hat{\phi}\left(\frac{w}{2}\right)$$

with $m_o(w/2)$ defined on $[-2\pi, 2\pi]$ as

$$m_o\left(\frac{w}{2}\right) = \begin{cases} \hat{\phi}(w) & |w| \le \frac{4\pi}{3} \\ 0 & \frac{4\pi}{3} < |w| \le 2\pi \end{cases}$$

and extended (4π) periodically to all $w \in \mathbb{R}$. The mother wavelet is given by

$$\hat{\psi}(w) = e^{-iw/2} \left[\int_{|\frac{w}{2}|-\pi}^{|w|-\pi} dP \right]^{\frac{1}{2}}.$$

Example 5 results from taking $dP(w) = \delta(w)dw$, while Example 6 results from taking P to be a C^∞ function with compact support in $[-\frac{\pi}{3}, \frac{\pi}{3}]$. Notice that the coefficients of the dilation equation are given by the Fourier series of $\hat{\phi}(2w)$

$$m_o(w) \sim \sum_k c_k e^{-iwk}, \quad c_k = \frac{1}{2\pi} \int_{-\pi}^{\pi} \hat{\phi}(2w) e^{iwk} dw.$$

As additional examples, we can use the rescaled B-splines of order n (Example 4) as our probability,

$$dP_n(w) = \theta_n \left((\frac{3}{\pi}w + 1)\frac{n+1}{2} \right) \frac{3}{2\pi}(n+1)dw, \quad n = 0, 1, 2, \cdots.$$

Since θ_n has support on $[0, n + 1]$, we have the desired support for P_n.

3.3   Reproducing kernels and a moment condition

Each of the spaces V_m in the multiresolution analysis is a reproducing kernel Hilbert space (RKHS) [Ar]. Such a space consists of a Hilbert space H of functions f on an interval T in which all evaluation functions $\xi_t(f) := f(t), f \in H, t \in T$ are continuous on H. Then by the Riesz representation theorem for each $t \in T$ there is a unique element $k_t \in H$ such that for each $f \in H$

$$f(t) = \langle f, k_t \rangle.$$

The function defined by $k(t, u) = \langle k_t, k_u \rangle$ for $t, w \in T$ is the reproducing kernel. Note that $L^2(T)$ is not a RKHS but has subspaces that are. The reproducing kernel of V_0 is given by

$$q(x, t) = \sum_n \overline{\phi(x - n)} \phi(t - n), \tag{3.17}$$

where $\phi(x)$ is the scaling function. The series and its derivatives with respect to t of order $\leq r$, converge uniformly for $x \in \mathbb{R}$, since, by the regularity of ϕ

$$|\phi^{(k)}(x)| \leq \frac{C_{kn}}{(1 + |x|)^n}, \quad k = 0, 1, \cdots, r; \ n = 0, 1, 2, \cdots.$$

We then use the inequality

$$(1 + |b|)(1 + |a|) \geq 1 + |b - a| \quad a, b \in \mathbb{R}$$

to find that

$$|q(x,t)| \le \sum_j |\phi(x-j)||\phi(t-j)|$$

$$\le C_{0,n+2}^2 \sum_j \frac{1}{(1+|x-j|)^{n+2}} \frac{1}{(1+|t-j|)^{n+2}}$$

$$\le C_{0,n+2}^2 \sum_j \frac{1}{(1+|x-j|)^2} \frac{1}{(1+|t-j|)^2} \frac{1}{(1+|x-t|)^n}$$

and hence that

$$|q(x,t)| \le \frac{C_n}{(1+|x-t|)^n}, \quad n \in \mathbb{N}.$$

The RK for V_m is obtained from that of V_0 by

$$q_m(x,t) = 2^m q(2^m x, 2^m t).$$

Similarly W_m, the orthogonal complements of V_m in V_{m+1} is a RKHS with RK

$$r_m(x,t) = 2^m \sum_n \overline{\psi(2^m x - n)} \psi(2^m t - n)$$

where $\psi(t)$ is the mother wavelet.

In Example 1, $q(x,t)$ is easy to find in closed form. It is just

$$q(x,t) = \phi(x - [t])$$

where $[t]$ is the largest integer $\le t$. In Example 5, it is

$$q(x,t) = \phi(x - t)$$

but in other cases has no obvious closed form.

The sequence $\{q_m(x,t)\}$ is a delta sequence in S'_r, i.e. $q_m(x,t) \to \delta(x-t)$. This follows from the fact that

$$\int_{-\infty}^{\infty} q_m(x,t)\theta(t)dt \to \theta(x) \quad \text{as} \quad m \to \infty$$

for each $\theta \in S_r$ where convergence is in the L^2 sense. As we shall see in a later chapter, this convergence is uniform on bounded sets. These kernels have a number of interesting properties, some of which come out of the wavelet moment theorem.

3.4   Regularity of wavelets as a moment condition

We have assumed that the scaling function ϕ and, hence, the mother wavelet $\psi \in S_r$. This condition leads to another condition in which ψ is orthogonal to all polynomials of degree $\le r$.

THEOREM 3.1

Let $\psi \in S_r$ with $\psi_{mn}(x) = 2^{m/2}\psi(2^m x - n)$ an orthonormal system in $L^2(\mathbb{R})$. Then

$$\int_{-\infty}^{\infty} x^k \psi(x)dx = 0, \qquad k = 0, 1, \cdots, r.$$

The proof is a straightforward application of the Lebesgue dominated convergence theorem and is similar to the one in [D]. We proceed by induction. Let N be a dyadic rational number $(= 2^{-j_o}k_o)$ such that $\psi(N) \neq 0$. This is possible since $\psi \neq 0$ and is continuous, and since such numbers are dense in $\mathbb{R}$. Let $j > 1$ be sufficiently large so that $2^j N$ is an integer. Then by orthogonality

$$0 = 2^j \int_{-\infty}^{\infty} \psi(x)\psi(2^j x - 2^j N)dx = \int_{-\infty}^{\infty} \psi(2^{-j}y + N)\psi(y)dy. \quad (3.18)$$

Since $\psi(2^{-j}y + N)$ is uniformly bounded, the integrand is dominated by a multiple of $|\psi(y)|$ and hence by the dominated convergence theorem

$$0 = \lim_{j \to \infty} \int_{-\infty}^{\infty} \psi(2^{-j}y + N)\psi(y)dy = \int_{-\infty}^{\infty} \lim_{j \to \infty} \psi(2^{-j}y + N)\psi(y)dy$$

$$= \int_{-\infty}^{\infty} \psi(N)\psi(y)dy = \psi(N) \int_{-\infty}^{\infty} \psi(y)dy.$$

Hence the conclusion holds for $k = 0$. By the induction hypothesis we assume it holds for $k = 0, 1, \cdots, n - 1$; $n \leq r$, and deduce, by Taylor's theorem,

$$\psi(x) = \sum_{k=0}^{n} \psi^{(k)}(N)\frac{(X - N)^k}{k!} + r_n(x)\frac{(X - N)^n}{n!},$$

where $r_n(x)$ is uniformly bounded and $\lim_{x \to N} r_n(x) = 0$. We now assume that $\psi^{(n)}(N) \neq 0$, and find by substituting this in (3.18) that

$$0 = \int_{-\infty}^{\infty} \left\{ \sum_{k=0}^{n} \psi^{(k)}(N)\frac{(2^{-j}y)^k}{k!} + r_n(2^{-j}y + N)\frac{(2^{-j}y)^n}{n!} \right\} \psi(y)dy$$

$$= \psi^{(n)}(N) \int_{-\infty}^{\infty} 2^{-jn}\frac{y^n}{n!}\psi(y)dy + \int_{-\infty}^{\infty} r_n(2^{-j}y + N)\frac{2^{-jn}y^n}{n!}\psi(y)dy.$$

We now multiply both sides by $2^{jn}n!$ and then take the limit as $j \to \infty$ to obtain

$$0 = \psi^{(n)}(N) \int_{-\infty}^{\infty} y^n \psi(y)dy + \int_{-\infty}^{\infty} \lim_{j \to \infty} r_n(2^{-j}y + N)y^n \psi(y)dy$$

$$= \psi^{(n)}(N) \int_{-\infty}^{\infty} y^n \psi(y)dy + 0,$$

which gives us our conclusion.   □

This enables us to derive a number of properties of the scaling function and the reproducing kernel.

THEOREM 3.2
Let $\phi \in S_r$ be an orthonormal scaling function with multiresolution analysis $\{V_m\}$; let $q_m(x,t)$ be the reproducing kernel of V_m. Then if $\hat{\phi}(0) \geq 0$,

(i) $\hat{\phi}(2\pi k) = \delta_{ok}, \quad p \in \mathbb{Z}$,

(ii) $\displaystyle\sum_n \phi(x-n) = 1, \quad x \in \mathbb{R}$,

(iii) $\displaystyle\int_{-\infty}^{\infty} q(x,y)dy = 1, \quad x \in \mathbb{R}$,

(iv) $\displaystyle\int_{-\infty}^{\infty} q_m(x,y)y^k\,dy = x^k, \quad x \in \mathbb{R},\ 0 \leq k \leq r,\ m \in \mathbb{Z}$.

PROOF Let f be the function such that $\hat{f}(w) = \chi_{[0,1]}(w)$. Then the projection of f onto V_m is given by

$$f_m(x) = \int_{-\infty}^{\infty} q_m(x,t)f(t)dt = \frac{1}{2\pi}\int_{-\infty}^{\infty} \hat{q}_m(x,w)\hat{f}(w)dw$$
$$= \frac{1}{2\pi}\int_0^1 \sum_n \hat{\phi}(w2^{-m})e^{-in2^{-m}w}\phi(2^m x - n)dw\,.$$

By Parseval's equality applied to $\{2^{m/2}\phi(2^m x - n)\}$ we have

$$\|f_m\|^2 = \frac{1}{(2\pi)^2}\sum_n \left| \int_0^1 \hat{\phi}(w2^{-m})e^{-iwn2^{-m}}dw \right|^2 2^{-m}\,.$$

Since $\left\{ e^{in2^{-m}w}/\sqrt{2^{m+1}\pi} \right\}$ is orthonormal on $[-2^m\pi, 2^m\pi]$ and the integral is its Fourier coefficients we again use Parseval's equality with respect to this system to obtain

$$\|f_m\|^2 = \frac{1}{2\pi}\int_0^1 |\hat{\phi}(w2^{-m})|^2 dw\,.$$

We now use Lebesgue's dominated convergence theorem and take the limit as $m \to \infty$. This gives us

$$\|f\|^2 = \frac{1}{2\pi}\int_0^1 |\hat{f}(w)|^2 dw = \frac{1}{2\pi} = \frac{1}{2\pi}\int_0^1 |\hat{\phi}(0)|^2 dw\,.$$

Thus $\hat{\phi}(0) = 1$, since we have required $\hat{\phi}(0) \geq 0$. The orthonormality of $\{\phi(t - n)\}$ as we have seen, translates into

$$\sum_k |\hat{\phi}(w + 2\pi k)|^2 \equiv 1,$$

from which it follows that $\hat{\phi}(2\pi k) = 0$, $k \neq 0$, thereby proving (i).

The second result is obtained by taking the Fourier series expansion of the periodic function

$$\sum_n \phi(x - n) = \sum_k c_k e^{i2\pi k x},$$

where

$$c_k = \int_0^1 \left(\sum_n \phi(x - n) \right) e^{-2\pi i k x} dx = \int_{-\infty}^\infty \phi(x) e^{-2\pi i k x} dx$$
$$= \hat{\phi}(2\pi k) = \delta_{ok}.$$

Hence we get (ii),

$$\sum_n \phi(x - n) = c_o = 1,$$

and (iii) is a corollary to (i) and (ii).

In order to prove (iv) we use Theorem 3.1. We first observe that

$$\int_{-\infty}^\infty \psi(y - k) y^n dy = 0, \qquad 0 \leq n \leq r$$

and use the fact that the reproducing kernel of V_1 satisfies

$$q_1(x, y) = q(x, y) + r(x, y),$$

where $r(x, y)$ is the reproducing kernel of W_o. Then it follows that

$$\int q_1(x, y) y^n dy = \int q(x, y) y^n dy + \int r(x, y) y^n dy,$$

and since

$$\int r(x, y) y^n dy = \int \sum_k \psi(x - k) \psi(y - k) y^n dy = 0,$$

that

$$\int q(x, y) y^n dy = \int q_1(x, y) y^n dy = \cdots = \int q_m(x, y) y^n dy.$$

We now write this last integral as

$$\int q_m(x, y)y^n(\theta(y) + (1 - \theta(y)))dy$$

where $\theta(y) \in S_r$, $0 \le \theta(y) \le 1$, and $\theta(y) = 1$, $|y| \le 2$. Then

$$\int q_m(x, y)y^n\theta(y)dy \to x^n\theta(y) = x^n, \quad |x| \le 1$$

since $q_m(x, y)$ is a delta sequence. On the other hand, we have

$$\int q_m(x, y)y^n(1 - \theta(y))dy$$

$$= \left\{\int_{-\infty}^{-2} + \int_2^{\infty}\right\} q_m(x, y)y^n(1 - \theta(y))dy,$$

and

$$\int_2^{\infty} |q_m(x, y)y^n(1 - \theta(y))|dy \le$$

$$\le \int_2^{\infty} \frac{C_{n+2}}{(1 + 2^m|x - y|)^{n+2}} y^n \, dy$$

$$\le \frac{C_{n+2}2^n}{(1 + 2^m(2 - 1))^n} \int_2^{\infty} \frac{1}{(1 + 2^m(y - 1))^2} dy$$

which converges to 0 as $m \to \infty$ uniformly for $|x| \le 1$. The same is true for the other integral.

The condition $|x| \le 1$ can be removed by a change of scale to deduce that

$$\int q_m(x, y)y^n dy \to x^n, \qquad \text{as } m \to \infty, \, x \in \mathbb{R}, \, n \le r.$$

Since the integral does not change with m, it follows that (iv) must hold.   □

3.5   Mallat's decomposition and reconstruction algorithm

The orthogonalities that arise together with the dilation equations enable us to construct a simple algorithm [Ma] relating the coefficients at different scales. For the space V_1, we have two distinct orthonormal bases:

$$\left\{\sqrt{2}\phi(2x - n)\right\}_{n=-\infty}^{\infty}$$

and

$$\{\phi(x - n), \psi(x - k)\}_{n,k=-\infty}^{\infty}.$$

Hence each $f \in V_1$ has an expansion

$$f(x) = \sum_n a_n^1 \sqrt{2}\phi(2x - n)$$
$$= \sum a_n^0 \phi(x - n) + \sum b_n^0 \psi(x - n).$$

By (3.3) and (3.16) we have

$$\phi(x - n) = \sqrt{2} \sum_k c_k \phi(2x - 2n - k),$$
$$\psi(x - n) = \sqrt{2} \sum_k (-1)^k c_{1-k} \phi(2x - n - k).$$

We substitute this into the expansion to obtain

$$f(x) = \sum_n a_n^0 \sqrt{2} \sum_k c_k \phi(2x - 2n - k)$$
$$+ \sum_n b_n^0 \sqrt{2} \sum_k (-1)^k c_{1-k} \phi(2x - 2n - k)$$
$$= \sum_n a_n^0 \sum_j c_{j-2n} \phi(2x - j)\sqrt{2}$$
$$+ \sum_k b_n^0 \sum_j (-1)^j c_{1-j+2n} \phi(2x - j)\sqrt{2}.$$

Hence by equating the coefficients, we have the algorithm

$$a_n^1 = \sum_k c_{n-2k} a_k^0 + \sum_k (-1)^n c_{1-n+2k} b_k^0, \qquad (3.19)$$

which is the reconstruction part. The decomposition is even easier: we need merely use the formula for a_n^0 and b_n^0 to find

(a) $$a_n^0 = \int_{-\infty}^{\infty} f(x)\phi(x - n)dx = \int_{-\infty}^{\infty} f(x) \sum_k c_k \sqrt{2}\phi(2x - 2n - k)dx$$
$$= \sum_k c_k a_{2n+k}^1 = \sum_k a_k^1 c_{k-2n}, \qquad (3.20)$$

(b) $$b_n^0 = \sum_k a_k^1 (-1)^k c_{1-k+2n}.$$

This works at each scale to give us the tree algorithm for decomposition (3.20),

$$
\begin{array}{ccccccccccc}
 & b_n^{m-1} & & b_n^{m-2} & & & & b_n^0 & & \\
 & \nearrow & & \nearrow & & \nearrow & & & \nearrow & & \nearrow \\
a_n^m & \longrightarrow & a_n^{m-1} & \longrightarrow & a_n^{m-2} & \longrightarrow & \cdots & a_n^1 & \longrightarrow & a_n^0 & \longrightarrow & \cdots,
\end{array}
$$

and for reconstruction (3.19)

$$
\begin{array}{ccccccccc}
& b_n^0 & & b_n^1 & & & & b_n^{m-1} & \\
& \searrow & & \searrow & & \searrow & & \searrow & \\
\cdots \longrightarrow & a_n^0 & \longrightarrow & a_n^1 & \longrightarrow & a_n^2 & \cdots \longrightarrow & a_n^{m-1} & \longrightarrow & a_n^m.
\end{array}
$$

Thus we calculate the coefficient from the function $f(t)$ only once at the finest scale of interest. Then we work down to successively coarser scales by using this decomposition algorithm, with the error at each successive scale corresponding to the wavelet coefficients.

3.6 Filters

A continuous linear system is a device which transforms a signal $x(t) \in L^2(\mathbb{R})$ linearly. It has an associated "impulse response" $h(t)$ such that the output of the system is given by the convolution,

$$
y = (h * x) \tag{3.21}
$$

This convolution exists if, say, x and $h \in L^1(\mathbb{R})$. However if x is somewhat restricted, more general impulse responses are possible. ([Pa, p.402])

A linear system is a <u>filter</u> if it passes signals of certain frequencies and attenuates others. This is best seen by taking the Fourier transform of (3.21) to get

$$
\hat{y}(w) = H(w)\hat{x}(w), \tag{3.22}
$$

where $H(w)$ is the system transfer function. If $H(w) = 0$, $|w| \geq w_0$, then H is a <u>low-pass filter</u> while on the other hand, if $H(w) = 0$ for $|w| \leq w_1$, then H is a <u>high-pass filter</u>. A <u>band-pass</u> filter passes a band $w_o \leq |w| \leq w_1$.

Similar concepts apply to discrete signals $x_n \in \ell^2$. Then we have the discrete impulse response (h_n)

$$
y_n = \sum_k h_{n-k} \chi_k.
$$

By taking Fourier series we find

$$
Y(w) = \sum y_n e^{iwn} = H(w)X(w),
$$

where each of these functions now is 2π periodic. The same concepts apply except everything is restricted to the interval $[-\pi, \pi]$. Thus, a low-pass filter would correspond to $H(w) = 0$ for $\frac{\pi}{2} \leq |w| \leq \pi$.

A number of operations associated with wavelets correspond to filters. The first is the projection onto the subspace V_m given by

$$f_m(t) := P_m f(t) = \int_{-\infty}^{\infty} q_m(t, y) f(y) dy.$$

For $m = 0$ this is given by

$$f_o(t) = \frac{1}{2\pi} \int \sum_n \phi(t - n) \overline{\hat{\phi}(w) e^{-iwn}} \, \hat{f}(w) dw$$

$$= \sum_n \phi(t - n) \frac{1}{2\pi} \int \overline{\hat{\phi}(w)} \, \hat{f}(w) e^{iwn} dw,$$

and hence

$$\hat{f}_o(\xi) = \sum_n \hat{\phi}(\xi) e^{-i\xi n}$$

$$\cdot \frac{1}{2\pi} \int_0^{2\pi} \sum_k \overline{\hat{\phi}(w + 2\pi k)} \, \hat{f}(w + 2\pi k) e^{iwn} dw \qquad (3.23)$$

$$= \hat{\phi}(\xi) \sum_k \overline{\hat{\phi}(\xi + 2\pi k)} \, \hat{f}(\xi + 2\pi k),$$

since the integral in (3.23) is just the Fourier coefficient of $\sum_k \overline{\hat{\phi}(\xi + 2\pi k)} \, \hat{f}(\xi + 2\pi k)$.

In the case of Meyer's wavelets $\hat{\phi}(\xi) = 0$ for $|\xi| > \pi + \epsilon$ and

$$\hat{f}_o(\xi) = |\hat{\phi}(\xi)|^2 f(\xi) + \hat{\phi}(\xi) \overline{\hat{\phi}(\xi + 2\pi)} \, f(\xi + 2\pi)$$

$$+ \hat{\phi}(\xi) \overline{\hat{\phi}(\xi - 2\pi)} \, f(\xi - 2\pi).$$

The last two terms have support on intervals of length $\leq 2\epsilon$ and can therefore be made arbitrarily small. Thus, the projection of f onto V_o is, except for this small term, a low-pass filter with system function $H(w) = |\hat{\phi}(w)|^2$.

The projection onto the subspaces W_m can, by similar arguments, be shown to be approximately a band-pass filter. For wavelets other than Meyer's, these conclusions are still approximately valid but with a poorer approximation.

The coefficient sequences are also related to discrete filters. The decomposition algorithm gives the coefficient at the coarser scale in terms of the finer scale. We have at $m = 0$, by (3.20)

$$a_n^0 = \sum_k c_{k-2n} a_k^1, \quad b_n^0 = \sum_k (-1)^k c_{1-k+2n} a_k^1.$$

These operations can be decomposed into a filter

$$e_n^0 = \sum_k c_{k-n} a_k^1$$

followed by decimation (take only even terms)

$$a_n^0 = e_{2n}^0,$$

and similarly for b_n. The filter has impulse response $\{c_{-n}\}$ since e_n^0 is the convolution of c_{-n} and a_k^1 and hence the system function is

$$H(w) = \sum_n c_{-n} e^{iwn}.$$

But we recall that

$$\hat{\phi}(w) = \sum_k c_k e^{-iwk/2} \frac{1}{\sqrt{2}} \hat{\phi}\left(\frac{w}{2}\right)$$

$$= m_o\left(\frac{w}{2}\right) \hat{\phi}\left(\frac{w}{2}\right),$$

i.e., the system function is exactly $H(w) = m_o(w)\sqrt{2}$.

In the Meyer case again, we have seen that

$$m_o\left(\frac{w}{2}\right) = \sum_k \hat{\phi}(w + 4\pi k),$$

and $H(w)$ therefore has support on $|w| < \frac{\pi \pm \epsilon}{2}$ in the interval $[-\pi, \pi]$. Therefore c_k is the impulse response of a discrete low-pass filter. For the wavelet coefficient, on the other hand, we get the system function for the corresponding filter

$$G(w) = \sum_k c_{k+1} e^{iwk} e^{i\pi k}$$

$$= \sum_n c_{-n} e^{-i(w+\pi)(n+1)}$$

$$= e^{-i(w+\pi)} H(-w - \pi).$$

Thus the support of $G(w)$ in $[-\pi, \pi]$ is contained in the set where $|w| \geq \frac{\pi}{2} - \frac{\epsilon}{2}$, and $d_k = (-1)^k c_{1-k}$ is the impulse response of a high-pass filter. The decomposition algorithm is given symbolically by

where the symbol $2\searrow$ denotes the removal of every other term in the sequence.

The reconstruction algorithm is treated similarly,

$$a_n^1 = \sum_k c_{n-2k} a_k^0 + \sum_k d_{n-2k} b_k^0.$$

This may be made into a pair of filters by interlacing 0's in the sequence $\{a_k^0\}$ and $\{b_k^0\}$. Then a_n^1 becomes

$$a_n^1 = \sum_j c_{n-j} a_j^{00} + \sum_j d_{n-j} b_j^{00},$$

where $\{a_j^{00}\}$ and $\{b_j^{00}\}$ are the sequences with the interlaced zeros. The first term corresponds to a low-pass and the second to a high-pass filter. It may be symbolized by

where the $2\nearrow$ denotes the operation of interlacing zeros. Note that the filters H and H^* and G and G^* have conjugate system functions, respectively.

3.7   Problems

1. Use the iterative procedure given in (3.11) to graph the Daubechies scaling functions (Example 3) for $\nu = 1/\sqrt{3}$. (You will need a computer for this; any program that handles iteration will work.)

2. Repeat Problem 1 for $\nu = \frac{1}{2}$ and compare.

3. Find a graph of the mother wavelet by using (3.16) in Problems 1 and 2.

4. Find an expression for the Fourier transform of the scaling function for the Battle-Lemarie wavelet based on B-splines of order 2 (Example 4) by repeating the orthogonalization procedure used in Example 2.

5. Use Gram-Schmidt orthogonalization on translates of the hat function (Example 2) $\{\theta(t-n)\}$. Is this an orthogonal basis of V_0? Is this a scaling function basis?

6. Let P be the uniform probability measure on $[-\epsilon, \epsilon]$, $0 < \epsilon < \frac{\pi}{3}$. Find the corresponding Meyer wavelet (Example 7) scaling function. Graph $\hat{\phi}(w)$, $m_o(w)$, and $|\hat{\psi}(w)|$.

7. Prove that

$$\int_{-\infty}^{\infty} t^m \psi(t)dt = 0, \quad m = 0, 1, 2, \cdots$$

for the mother wavelet $\psi(t)$ of the previous problem.

8. Show $\hat{\phi}(2\pi k) = \delta_{0k}$ for Problem 6 and for Problem 4.

9. Let the expansion coefficients in V_m, a_n^m for the scaling function of Problem 2 be given by $a_n^m = 2^{-n}$. Find a_n^{m-1} and b_n^{m-1} by using (3.20).

10. Check Problem 9 by using the reconstruction algorithm (3.19).

11. Find the Fourier transform of a projection of f onto V_0 in the case of Problem 6. Find an expression for the difference between this and the ideal low-pass filter.

12. Repeat Problem 11 for W_0 and a band-pass filter.

4

Convergence and Summability
of Fourier Series

In this chapter we extend and refine the introduction to trigonometric Fourier series begun in Chapter 1. In addition to the pointwise convergence results presented there, we consider stronger results and rates of convergence. These other results include Gibbs' phenomenon and two types of summability: Cesàro and Abel. We then consider periodic tempered distributions and use them for a more general theory of Fourier series in which the distinction between formal trigonometric series and Fourier series is eliminated.

4.1 Pointwise convergence

There are many tests for the pointwise convergence of Fourier series [Z]. The one given in Chapter 1 is quite weak and excludes many functions for which we have such convergence. We present only one involving Lipschitz continuity.

Let $f(x)$ be a 2π periodic function whose restriction to $(-\pi, \pi)$ is in $L^1(-\pi, \pi)$; f is said to satisfy a <u>Lipschitz condition</u> of order $\alpha > 0$ at x_0 if there is a constant C such that

$$|f(x) - f(x_0)| \leq C|x - x_0|^\alpha \tag{4.1}$$

in some neighborhood of x_0. If the condition holds for all x_0 with the same C then f satisfies a <u>uniform Lipschitz</u> condition. We restrict α to be ≤ 1 since the condition implies that $f'(x) = 0$ for $\alpha > 1$. We first prove the Riemann-Lebesgue lemma.

PROPOSITION 4.1

Let $f \in L^1(-\pi, \pi)$; then the Fourier coefficients satisfy

$$|a_n| + |b_n| \to 0 \qquad \text{as } n \to \infty.$$

We already know this is true for $f \in L^2(-\pi, \pi)$ by Bessel's inequality (1.3). Since each $f \in L^1(-\pi, \pi)$ may be approximated by some $g \in L^2(-\pi, \pi)$ in the sense of $L^1(-\pi, \pi)$ the result follows.

PROPOSITION 4.2

Let f satisfy (4.1) at x_0; then

$$S_n(x_0) \to f(x_0)$$

as $n \to \infty$.

PROOF   The proof starts with formula (1.9), for $x_0 \neq \pm\pi$,

$$
\begin{aligned}
S_n(x_0) - f(x_0) &= \int_{-\pi}^{\pi} \left\{ \frac{f(x_0 - u) - f(x_0)}{u} \right\} u D_n(u) du \\
&\leq \int_{-\pi}^{-\delta} + \int_{-\delta}^{\delta} + \int_{\delta}^{\pi} = I_1 + I_2 + I_3.
\end{aligned}
\tag{4.2}
$$

The first integral can be written as

$$
\begin{aligned}
I_1 &= \int_{-\pi}^{-\delta} \left\{ \frac{f(x_0 - u) - f(x_0)}{u} \right\} \frac{u}{2\pi \sin u/2} \sin(n + \tfrac{1}{2}) u \, du \\
&= \frac{1}{\pi} \int_{-\pi}^{-\delta} \left\{ \frac{f(x_0 - u) - f(x_0)}{u} \right\} \frac{u/2}{\sin u/2} (\sin nu \cos u/2 \\
&\qquad + \cos nu \sin u/2) du \\
&= \frac{1}{\pi} \int_{-\pi}^{\pi} g(x_0, u) \sin nu \, du + \frac{1}{\pi} \int_{-\pi}^{\pi} h(x_0, u) \cos nu \, du \,,
\end{aligned}
$$

where g and h are both L^1 functions. Hence by proposition 4.1, $I_1 \to 0$ as $n \to \infty$. A similar argument holds for I_3, while I_2 satisfies

$$|I_2| \leq \frac{1}{2} \int_{-\delta}^{\delta} C|u|^{\alpha-1} du = \frac{C}{\alpha} \delta^{\alpha}.$$

Thus, given $\epsilon > 0$, we may choose δ such that $\frac{C}{\alpha}\delta^{\alpha} < \frac{\epsilon}{2}$ and then choose N such that $|I_1 + I_3| < \frac{\epsilon}{2}$ for $n \geq N$. This makes

$$|S_n(x_0) - f(x_0)| < \epsilon, \qquad n \geq N. \qquad \square$$

For $x_0 = \pm\pi$ we translate everything by an amount π and show that $S_n(x - \pi) \to f(x - \pi)$ at $x = 0$.   □

In order to get uniform convergence we need a uniform version of the Riemann-Lebesgue lemma.

LEMMA 4.3
Let f be periodic and in $L^1[-\pi, \pi]$ and let $h \in C^1[\alpha, \beta]$ where $[\alpha, \beta] \subseteq [-\pi, \pi]$; then

$$\int_\alpha^\beta f(x - u)h(u)\sin \lambda u \, du \to 0 \qquad \text{as } \lambda \to \infty$$

uniformly in x.

PROOF   We first approximate f by a periodic function $g \in C^1$ in the sense of $L^1(-\pi, \pi)$, i.e., choose g such that

$$\int_{-\pi}^\pi |f(x) - g(x)| dx < \epsilon.$$

Now, we use integration by parts on

$$I_1 = \int_\alpha^\beta g(x - u)h(u)\sin d\mu \, du = -g(x - u)h(u)\frac{\cos \lambda u}{\lambda}\Big|_\alpha^\beta$$
$$+ \int_\alpha^\beta [g(x - u)h(u)]'\frac{\cos \lambda u}{\lambda} du.$$

Since both $g(x - u)$ and its derivative are uniformly bounded in x, this integral converges to 0 uniformly as $\lambda \to \infty$. Then we see that

$$\left| \int_\alpha^\beta f(x - u)h(u)\sin \lambda u \, du \right|$$
$$\leq \left| \int_\alpha^\beta (f(x - u) - g(x - u))h(u)\sin \lambda u \, du \right| + |I_1|$$
$$\leq \max_{\alpha \leq u \leq \beta} |h(u)| \int_\alpha^\beta |f(x - u) - g(x - u)| du + |I_1|$$
$$\leq \max_{\alpha \leq u \leq \beta} |h(u)| \epsilon + |I_1|.$$

The first term can be made arbitrarily small while the second converges to 0 as $\lambda \to \infty$.   □

As a corollary we obtain the following.

THEOREM 4.4
Let f satisfy a uniform Lipschitz condition of order $\alpha > 0$ in (a, b). Then $S_n \to f$ uniformly in any interior subinterval $[c, d] \subset (a, b)$.

PROOF Choose $\delta < \min(c - a, b - d)$. Then the integral I_2 in (4.2) satisfies $|I_2| \leq \frac{C}{\alpha} \delta^\alpha$ as before. For I_1 we use the lemma applied to $\alpha = -\pi$, $\beta = -\delta$ and take $\lambda = \left(n + \frac{1}{2}\right)$ to deduce that

$$\frac{1}{\pi} \int_{-\pi}^{-\delta} f(x_0 - u) \frac{1}{2 \sin u/2} \sin(n + \frac{1}{2})u \, du \to 0$$

uniformly as $n \to \infty$. Since the other term in I_1 clearly converges to 0 uniformly, it follows that I_1 does as well and similarly for I_2. $\square$

COROLLARY 4.5 (Localization principle)
Let f_1 and f_2 be equal in an interval (a, b); then $S_n(f_1)$ and $S_n(f_2)$ are uniformly equiconvergent in interior subintervals of (a, b).

The proof is immediate since $f_1 - f_2$ satisfies the hypothesis of Theorem 4.4. This says that the local convergence of the Fourier series depends only on the local behavior of the function.

Another extension is to functions that satisfy one sided Lipschitz conditions at x_0. Again the same formula leads to the conclusion that

$$S_n(x_0) \to \frac{f(x_0^+) + f(x_0^-)}{2}. \tag{4.3}$$

We cannot extend the above results from Lipschitz to ordinary continuity since there are many examples of continuous functions whose Fourier series diverge. [Z, p. 298]. In order to treat these, various types of summability have been introduced, which we will cover in the next section. Many of the results in convergence and summability arise because of the properties of certain delta sequences, i.e., sequences that converge to δ in S'. For ordinary convergence the associated delta sequence is the Dirichlet kernel $D_n(t)$ restricted to $(-\pi, \pi)$. It is a delta sequence since by Proposition 4.2,

$$\int_{-\pi}^{\pi} D_n(t)\phi(t)dt = S_n(0) \to \phi(0) = (\delta, \phi)$$

for any test function $\phi \in S$. That is, $\chi D_n \to \delta$ as $n \to \infty$ where χ is the characteristic function of $[-\pi, \pi]$. If we omit the χ, then D_n, being periodic, must converge to a periodic distribution, which we denote as $\delta^*(u)$. Clearly, $D_n \to \delta^*$ in S'.

The rate of convergence of the Fourier series can be studied by finding the rate of convergence of $\{D_n\}$ in the Sobolev space of periodic functions, $H_{2\pi}^\alpha$. (Compare Section 2.5.) It consists of all periodic $f \in S'$

such that

$$\sum_{n=-\infty}^{\infty} |c_n|^2 (n^2+1)^\alpha < \infty \qquad (4.4)$$

where c_n are the complex Fourier coefficients of f. For $\alpha = 0$ this is just $L^2(-\pi, \pi)$ while for $\alpha = 1, 2, \cdots$, it consists of f whose αth derivative is in $L^2(-\pi, \pi)$. For negative α, this involves the Fourier series of a periodic distribution, which we take up later in this chapter. Here we merely note that δ^*, the periodic extension of δ, has Fourier series

$$\delta^*(x) \sim \frac{1}{2\pi} \sum_{n=-\infty}^{\infty} e^{inx} \qquad (4.5)$$

and therefore $\delta^* \in H_{2\pi}^{-\alpha}$, $\alpha > \frac{1}{2}$. The norm in $H_{2\pi}^\alpha$ is given by the square root of (4.4) and the inner product by

$$\langle f, g \rangle_\alpha := \sum_n c_n \overline{d_n} (n^2+1)^\alpha.$$

It is clear that $D_n(t) \to \delta^*(t)$ in the sense of $H_{2\pi}^{-\alpha}$, $\alpha > \frac{1}{2}$. We can find the rate at which it converges, and can use this to find the rate of convergence of Fourier series. It is, since

$$D_n(x) - \delta^*(x) \sim \frac{1}{2\pi} \sum_{|k|>n} e^{ikx},$$

given by

$$\|D_n - \delta^*\|_{-\alpha}^2 = \left(\frac{1}{2\pi}\right)^2 \sum_{|k|>n} (k^2+1)^{-\alpha} \le 2 \left(\frac{1}{2\pi}\right)^2 \int_{n+1}^{\infty} t^{-2\alpha} \, dt$$

$$= \frac{1}{2\pi^2} \frac{(n+1)^{-2\alpha+1}}{2\alpha-1},$$

which we restate as

PROPOSITION 4.6
The Dirichlet kernel $\{D_n\}$ converges to δ^* in $H_{2\pi}^{-\alpha}$ at a rate $O\left(n^{-\alpha+\frac{1}{2}}\right)$ for $\alpha > \frac{1}{2}$.

COROLLARY 4.7
Let $f \in H_{2\pi}^\beta$, $\beta > \frac{1}{2}$; then the Fourier series of f converges to f uniformly at a rate $O\left(n^{-\beta+\frac{1}{2}}\right)$ for $\beta > \frac{1}{2}$.

The proof involves merely using the Sobolev version of Schwarz's inequality, viz. if $f \in H_{2\pi}^{-\alpha}$ and $g \in H_{2\pi}^{\alpha}$

$$|(f,g)| \leq \|f\|_{2\pi}^{-\alpha} \|g\|_{2\pi}^{\alpha}$$

where (f,g) denotes the value of the functional f on the test function g. (See Chapter 2 where the test functions however were not periodic.) Since the partial sums of the Fourier series of f are given by

$$f_n(x) = (D_n * f)(x),$$

it follows that

$$|f_n(x) - f(x)| \leq \|D_n - \delta^*\|_{-\beta} \|f\|_{\beta}. \tag{4.6}$$

4.2 Summability

As we have observed, Fourier series do not necessarily converge pointwise for functions that are continuous. However, by replacing the partial sums of the series by other sums, we can get uniform convergence for all continuous functions.

Let $f(x)$ be a continuous periodic function whose Fourier series has partial sums $S_n(x)$, then the C-1 (Cesàro) means of the Fourier series are

$$\sigma_n(x) := \frac{1}{n} \sum_{k=0}^{n-1} S_k(x).$$

Just as with the partial sums, $\sigma_n(x)$ may be expressed as an integral involving $f(x)$

$$\sigma_n(x) = \frac{1}{n} \sum_{k=0}^{n-1} (D_k * f)(x)$$

$$= \int_{-\pi}^{\pi} \frac{1}{n} \sum_{k=0}^{n-1} D_k(x-t) f(t) dt = (F_n * f)(x), \tag{4.7}$$

where the kernel $F_n(x) := \frac{1}{n} \sum_{k=0}^{n-1} D_k(x)$ may be evaluated by the same trick used before

$$\sin^2 \frac{1}{2} x F_n(x) = \frac{1}{n} \sum_{k=0}^{n-1} \frac{\sin \frac{1}{2} x \sin \left(k + \frac{1}{2}\right) x}{2\pi}$$

$$= \frac{1}{2\pi n} \sum_{k=0}^{n-1} \frac{\cos kx - \cos(k+1)x}{2} \tag{4.8}$$

$$= \frac{1}{2\pi n} \left(\frac{1}{2} - \frac{\cos nx}{2} \right) = \frac{1}{2\pi n} \sin^2 \frac{nx}{2}.$$

This is usually denoted the Fejer kernel and is easily shown to be a delta sequence as in the Dirichlet kernel D_n. However, it belongs to a class of such sequences, the quasi-positive delta sequences to which D_n does not belong. We shall discuss these in more detail in Chapter 8. The properties of χF_n include the following:

(i) $\displaystyle\int_{-\pi}^{\pi} F_n(x)dx = 1, \qquad n \in \mathbb{N},$

(ii) $F_n(x) \geq 0, \qquad x \in \mathbb{R}, \quad n \in \mathbb{N},$ (4.9)

(iii)   For each $\gamma > 0$, $\displaystyle\sup_{\pi > |x| > \gamma} F_n(x)\chi_\pi(x) \to 0,$ as $n \to \infty.$

This is actually a positive delta sequence, which is stronger than quasi-positive (see Chapter 8), which, however, is all that is needed for uniform convergence. The proof that the C-1 means converge uniformly is considerably simpler. (See, e.g., [Z, p.87].)

Another type of summability, Abel summability, leads to a positive delta family which depends on a parameter $0 < r < 1$. The Abel means σ_r are given by

$$\sigma_r(x) := \sum_{n=-\infty}^{\infty} c_n r^{|n|} e^{inx} \qquad 0 < r < 1. \tag{4.10}$$

This series clearly converges uniformly in x for each such r. The kernel, the so-called Poisson kernel, is

$$P_r(x) := \frac{1}{2\pi} \sum_{n=-\infty}^{\infty} r^{|n|} e^{inx} = \frac{1 - r^2}{2\pi(1 - 2r\cos x + r^2)}. \tag{4.11}$$

It satisfies the same conditions as (4.9) except that $n \to \infty$ is replaced by $r \to 1$. Since $\sigma_r = P_r * f$, it again follows that the $\sigma_r \to f$ uniformly for each continuous f.

4.3   Gibbs' phenomenon

There seems to be some misunderstanding as to what constitutes Gibbs' phenomenon. It deals not with the failure of the Fourier series to converge at a point of jump discontinuity but rather with the "overshoot" of the partial sums in the limit.

In order to demonstrate Gibbs' phenomenon, we consider the behavior of a simple function with a jump discontinuity at 0, $f(x) = \frac{\pi}{2} \operatorname{sgn} x - \frac{x}{2}$,

$|x| < \pi$. The Fourier series of $f(x)$ is

$$f(x) = \sum_{n=1}^{\infty} \frac{\sin nx}{n} = \sum_{n=1}^{\infty} \int_0^x \cos nt \, dt$$

$$= \lim_{n \to \infty} \int_0^x \left(\frac{1}{2} + \sum_{k=1}^n \cos kt \right) dt - \frac{x}{2} \qquad (4.12)$$

$$= \lim_{n \to \infty} \left(\pi \int_0^x D_n(t) dt \right) - \frac{x}{2}.$$

Since f satisfies a left and right Lipschitz condition at 0, and the average of the left and right hand values of f at 0 is 0, the series converges to 0 at 0. We wish, however, to investigate the partial sums of the series as $x \to 0^+$. The integral in (4.12) may be written as

$$\pi \int_0^x D_n(t) dt = \int_0^x \left(\frac{\sin nt \cos \frac{1}{2}t}{2 \sin \frac{1}{2}t} + \cos nt \right) dt$$

$$= \int_0^x \frac{\sin nt}{t} dt + \int_0^x \sin nt \left(\frac{\cos t/2}{2 \sin t/2} - \frac{1}{t} \right) dt \qquad (4.13)$$

$$+ \int_0^x \cos nt \, dt$$

for $x > 0$. The second and third integral converge uniformly to 0 by the Riemann-Lebesgue lemma (or by direct calculation); that leaves only the first integral, which may be written as

$$I(nx) = \int_0^{nx} \frac{\sin t}{t} dt \to \int_0^{\infty} \frac{\sin t}{t} dt = \frac{\pi}{2}, \quad \text{for fixed } x > 0.$$

But if we take the sequence $x_n = \frac{\pi}{n}$, the integral becomes

$$I(nx_n) = \int_0^{\pi} \frac{\sin t}{t} dt > \frac{\pi}{2}.$$

In fact the ratio between the two integrals is almost 1.18 [Z, p.61]. Hence the partial sums of the original function at x_n are given by

$$S_n(x_n) = I(nx_n) - \frac{x_n}{2} + \epsilon_n,$$

which converges to a value $> f(0^+)$ as $n \to \infty$. Such an overshoot appears whenever the function has a jump discontinuity [Z, p.61].

4.4   Periodic distributions

In Chapter 2, we mentioned periodic distributions and their Fourier series briefly. Here we enlarge on these results. The coefficients, given by (2.6), are independent of t since their derivatives are zero. Since each $f \in S'$ may be expressed as

$$f = D^p F ,$$

where F is a continuous function of polynomial growth, (2.6) may be written as

$$c_n = \frac{1}{2\pi} \left[T_{2\pi}(D^p F e^{-int})^{(-1)} - (D^p F e^{-int})^{(-1)} \right] . \qquad (4.14)$$

Since f is periodic it follows that $T_{2\pi} F - F$ is a polynomial R of degree less than p. We use this and integration by parts to get a bound on c_n. Indeed

$$(D^p F e^{-int})^{(-1)} = \sum_{j=0}^{p-1} D^j F(in)^{p-1-j} e^{-int} + (in)^p \left(F(t) e^{-int}\right)^{(-1)}$$

and hence c_n is given by

$$c_n = \frac{1}{2\pi} \left[\sum_{j=0}^{p-1} D^j R(in)^{p-1-j} e^{-int} + (in)^p \int_t^{t+2\pi} F(x) e^{-inx} dx \right]$$

from which it follows that

$$|c_n| \leq \frac{1}{2\pi} \sum_{j=0}^{p} c_j(t) |n|^j , \quad |n| \geq 1,$$

where $c_j(t)$ is bounded on compact sets. Hence since c_n is constant, $c_n = 0(|n|^p)$.

The converse is also true; each series $\sum c_n e^{int}$ where $c_n = 0(|n|^p)$ converges in S' to a periodic distribution. This is fairly easy to show since

$$\sum_n c_n(e^{int}, \phi) = \sum_n c_n \hat{\phi}(-n) \qquad (4.15)$$

for $\phi \in S$. But $\hat{\phi} \in S$ also and hence is rapidly decreasing so that the series (4.15) converges. Since convergence in S' is weak convergence, the limit must be an element $f \in S'$.

We have proved parts of

THEOREM 4.8
Let $\{c_n\}_{n=-\infty}^{\infty}$ *be a sequence of complex numbers; then the following are equivalent:*
  (i)   *there is a* $p \in \mathbb{Z}$ *such that* $c_n = 0(|n|^p)$,
  (ii) $\sum_n c_n e^{int}$ *converges in the sense of* S',
  (iii) *the* c_n *are the Fourier coefficients of some* $f \in S'$.

PROOF We have shown that (i) $\Rightarrow$ (ii) and (iii) $\Rightarrow$ (i). Thus it remains to show (ii) $\Rightarrow$ (iii). We first observe that the operations of translation and multiplication by e^{int} are continuous in S'. Antidifferentiation is also continuous in the sense that for some constants a_n, $f_n \to f$ in S' implies that $f_n^{(-1)} + a_n \to f^{(-1)}$ in S'. Thus, the Fourier coefficients of

$$f(t) = \sum_n c_n e^{int} = \lim_{N \to \infty} \sum_{|n| \leq N} c_n e^{int}$$

are given by

$$c_n' = \frac{1}{2\pi} \int_t^{t+2\pi} f(t) e^{-int} dt \,,$$

which is to be interpreted as (2.6). Because of the continuity of the various operations

$$c_n' = \lim_{N \to \infty} \frac{1}{2\pi} \int_t^{t+2\pi} \sum_{|k| \leq N} c_k e^{ikx} e^{-inx} dx$$

$$= \lim_{N \to \infty} \begin{cases} c_n, & n \leq N \\ 0, & n > N \end{cases} = c_n. \qquad \square$$

There is still one property to check in order to eliminate the distinction between a trigonometric and a Fourier series. If (iii) is satisfied, then by (ii), its Fourier series must converge to some $g \in S'$, which may not be the same as the f we started with.

PROPOSITION 4.9
Let f *be a periodic distribution whose Fourier coefficients are all zero; then* $f = 0$.

PROOF Since $c_0 = 0$, f has a periodic antiderivative. Moreover, its p-th order antiderivative F, the continuous function in (4.14), may be taken to be periodic. If not, we could add a polynomial P of degree $\leq p - 1$ and observe that

$$T_{2\pi}(F + P) - (F + P) = T_{2\pi}F - F + T_{2\pi}P - P = R + T_{2\pi}P - P.$$

Now R is a polynomial of degree $\leq p - 2$ since $T_{2\pi} F^{(p-1)} - F^{(p-1)}$ is the periodic antiderivative of f. Hence P may be chosen such that the polynomial

$$T_{2\pi} P - P$$

of degree $\leq p - 2$ is equal to $-R$.

If F is periodic, then the coefficients of f are given by

$$c_n = \frac{(in)^p}{2\pi} \int_t^{t+2\pi} F(x) e^{-inx} \, dx.$$

Thus, the Fourier coefficients of F are all zero except possibly for the 0-th coefficient. This implies that F is a constant and if $p > 0$, $f = D^p F = 0$. If $p = 0$, f is an ordinary continuous function, and the result is also true.   □

Since the periodic distribution $f \in S'$, it must have a Fourier transform, and the transform of the series must converge to it:

$$\hat{f}(w) = \sum_n c_n \widehat{(e^{int})}(w) = 2\pi \sum_n c_n \delta(w - n). \qquad (4.16)$$

Thus the support of $\hat{f}$ must be on the set of integers. In particular for $f = \delta^*$, we have

$$\delta^*(t) = \frac{1}{2\pi} \sum_n e^{int},$$

and hence

$$\hat{\delta}^*(w) = \frac{2\pi}{2\pi} \sum_n \delta(w - n) = \sum_n 2\pi \, \delta(2\pi w - 2\pi n) = 2\pi \, \delta^*(2\pi w). \quad (4.17)$$

If we now take the value of both sides on a test function $\phi \in S$ we obtain

$$(\hat{\delta}^*, \phi) = \sum_n (\delta(\cdot - n), \phi) = \sum_n \phi(n)$$

$$= (\delta^*, \hat{\phi}) = \sum_k \hat{\phi}(2\pi k).$$

This result is one form of the Poisson summation formula.

4.5   Problems

1. For which α does $f(x) = |x|^{1/2}$ satisfy a uniform Lipschitz condition of order α on $(-\pi, \pi)$?

2. Let $a_n \to 0$ and $a_1 \geq a_2 \geq \cdots \geq a_n \geq 0$. Show the series $\sum_{n=1}^{\infty} a_n \sin nx$ converges for all x. (Hint: Use the summation by parts formula

$$\sum_{n=1}^{N} a_n u_n = \sum_{n=1}^{N-1} (a_n - a_{n+1})U_n + a_N U_N,$$

where $U_n = \sum_{k=1}^{n} u_n$, and observe that for $u_n = \sin nx$, U_n is bounded.

3. Let $\{a_n\}$ be as in Problem 2. Show that

$$\sum_{n=1}^{\infty} a_n \cos nx \quad \text{converges for} \quad x \in (0, 2\pi).$$

4. Take $a_n = n^{-1/2}$ in Problem 2. Why doesn't the result of Problem 2 contradict Bessel's inequality?

5. Find the Fourier series of $f(x) = \chi_{[-1,1]}(x)$ on $(-\pi, \pi)$. Show that none of the even terms in the series are zero at $x = \frac{\pi}{2}$. Does this contradict the localization principle?

6. Find the C-1 means of the Fourier series of δ^* (4.5) and show they converge to 0 for $x \neq 0$, $x \in (-\pi, \pi)$.

7. Find the Abel means of the Fourier series of $\delta^{*\prime}$ and show they converge to zero for $x \neq 0$ $x \in (-\pi, \pi)$ as $r \to 1$.

8. Let $f(x)$ be the jump function whose Fourier series is given by (4.12). Let $\sigma_n(x)$ denote the C-1 means of $f(x)$. Show that for this function,

$$\sigma_n(x) = \pi \int_0^x F_n(t)dt - \frac{x}{2}, \quad 0 < x < \pi.$$

Use this to show that $|\sigma_n(x)| \leq \frac{\pi}{2}$ and hence Gibbs' phenomenon fails to hold.

9. Let $f(x)$ be as in Problem 8. Show that $\delta^*(x) - \frac{1}{2} = Df(x)$. Use this to find a continuous second order antiderivative of $\delta^*(x)$.

10. Let $f(x)$ be a function on $\mathbb{R}$ such that $f(x) = O((1+|x|)^{-1-\epsilon})$ and $\hat{f}(w) = O((1 + |w|)^{-1-\epsilon})$ for some $\epsilon > 0$. Show that

$\hat{f}*(w) = \sum\limits_{k=-\infty}^{\infty} \hat{f}(w + 2\pi k) \in L^1(-\pi, \pi)$ and find an expression for its Fourier series.

11. Use the result in Problem 10 to prove a general form of the Poisson summation formula

$$\sum_{k} \hat{f}(w + 2\pi k) = \sum_{n} f(-n)e^{iwn}.$$

12. Repeat Problem 11 but with $f(x)$ instead of $\hat{f}(w)$ to show that

$$\sum_{n} f(x - n) = \sum_{k} \hat{f}(2\pi k)e^{2\pi ikx}.$$

5

Wavelets and Tempered Distributions

Just as with other orthogonal systems, it is possible to extend the expansions in orthogonal wavelets from $L^2(\mathbb{R})$ to a certain class of tempered distributions. Since we are supposing in general that the scaling function ϕ is r-regular, i.e., $\phi \in S_r$, we might expect to obtain expansions for $f \in S'_r$ as indeed we do. Since the mother wavelet $\psi \in S_r$ if ϕ is, the expansion coefficients with respect to both $\{\phi(t-n)\}$ and $\{\psi(t-n)\}$ are well defined. In fact, since $f \in S'_r$ is characterized by

$$f = D^r\{(t^2+1)^k\mu\},$$

where μ is a bounded measure on $\mathbb{R}$, r, $k \in \mathbb{Z}$, the coefficients may be found to satisfy

$$
\begin{aligned}
a_n &= (f, \phi(\cdot - n)) \\
&= (D^r\{(t^2+1)^k\mu\}, \phi(\cdot - n)) \\
&= \int_{-\infty}^{\infty} (t^2+1)^k(-1)^r\phi^{(r)}(t-n)d\mu.
\end{aligned}
\tag{5.1}
$$

Since $(t^2+1)^k \leq (|t|+1)^{2k}$ and since $|t|+1 \leq (|n|+1)(|t-n|+1)$ it follows that

$$|a_n| \leq \int_{-\infty}^{\infty}(|n|+1)^{2k}(|t-n|+1)^{2k}|\phi^{(r)}(t-n)|d|\mu| = O(|n|^{2k}).$$

Similarly $b_n = (f, \psi(t-n))$ satisfies the same sort of growth condition.

The coefficients of $\theta \in S_r$ can similarly be calculated as

$$\int_{-\infty}^{\infty} \theta(t)\phi(t-n)dt = O(|n|)^{-p}), \qquad \text{for all } p \in \mathbb{Z}.$$

Thus the expansion of f converges in the sense of S'_r to some $f_0 \in S'_r$, i.e.,

$$\sum_n a_n \int \phi(t-n)\theta(t)dt = (f_0, \theta).$$

But the convergence of the expansion of f is much stronger than this since $\sum a_n \phi(t-n)$ converges uniformly on bounded sets as do its first r derivatives. This follows from the inequality

$$\sum_n |a_n \phi^{(j)}(t-n)| \le C \sum_n \frac{(|n|+1)^k}{(|t-n|+1)^{k+2}}$$

$$\le C \sum_n \sum_m \binom{k}{m} \frac{(|n-t|+1)^{k-m}|t|^m}{(|n-t|+1)^{k+2}} \qquad (5.2)$$

$$\le C \sum_n \frac{1}{(|n-t|+1)^2}(1+|t|)^k$$

$$\le C'(1+|t|)^k.$$

In fact we have shown that the limiting function and its r derivatives are continuous functions of polynomial growth on $\mathbb{R}$. These results enable us to imitate the multiresolution analysis of $L^2(\mathbb{R})$ in S'_r.

We can also turn things around and try to expand functions in wavelets composed of distributions. Rather than requiring $\phi \in S_r$, we suppose merely that $\phi \in S'$. Then, of course, we cannot orthogonalize ϕ in the sense of $L^2(\mathbb{R})$ but must use another Hilbert space. An appropriate one that works in certain cases is the Sobolev space H^{-s}. One scaling distribution we consider will be composed of linear combinations of delta functions. In this case, we again obtain a version of a multiresolution analysis. In the more general case, where we consider distribution solution of dilation equations, we are unable to obtain an orthogonal multiresolution and so only have partial results.

5.1 Multiresolution analysis of tempered distributions

DEFINITION 5.1   *Let $\{a_n\}$ be a sequence of complex numbers such that $a_n = O(|n|^k)$ for some $k \in \mathbb{Z}$; then $T_0 := \{f | f(t) = \sum_n a_n \phi(t-n)\}$ and $U_0 := \{g | g(t) = \sum_n a_n \psi(t-n)\}$.*

Hence both T_0 and U_0 are composed of functions in C^r of polynomial growth and are therefore in S'_r. We denote by T_m and U_m their corresponding dilation spaces ($f \in T_0 \Leftrightarrow f(2^m t) \in T_m$). Then we may expect that a multiresolution analysis of S'_r exists, namely,

$$\cdots \subset T_{-1} \subset T_0 \subset T_1 \subset \cdots \subset T_m \subset \cdots S'_r \qquad (5.3)$$

and

$$\overline{\bigcup_m T_m} = \mathcal{S}'_r,$$

where the closure is in the topology of $\mathcal{S}'_r$, but we must check a few things. Since $\phi(t) \in V_1 \subset T_1$ where V_1 is part of the $L^2(\mathbb{R})$ multiresolution analysis, it follows that all partial sums $\sum_{|n| \leq N} a_n \phi(t-n) \in T_1$.

However, the corresponding series may not be in T_1 unless $\{c_k\}$ in (3.3) is sufficiently rapidly decreasing. We therefore assume $c_k = O(|k|^{-m})$ for all $m \in \mathbb{Z}$, i.e. $\{c_k\}$ is rapidly decreasing. Then the series becomes

$$
\begin{aligned}
\sum_n a_n \phi(t-n) &= \sum_n a_n \sum_k \sqrt{2} c_k \phi(2t - 2n - k) \\
&= \sum_n a_n \sum_j c_{j-2n} \sqrt{2} \phi(2t - j) \\
&= \sum_j \left(\sum_n a_n c_{j-2n} \right) \sqrt{2} \phi(2t - j) \\
&= \sum_j a_j^1 \sqrt{2} \phi(2t - j).
\end{aligned}
\tag{5.4}
$$

Clearly if a_n is of polynomial growth $(= O(|n|^p))$, so is a_n^1.

It can also be shown that T_0 is closed in the sense of $\mathcal{S}'_r$. This follows from the fact that $f_m \to 0$ in $\mathcal{S}'_r$ is equivalent to

$$f_m = D^r [(1+t^2)^p \mu_m] \quad \text{and} \quad \int d|\mu_m| \to 0 \quad [\text{Ze}],$$

where $\{\mu_m\}$ is a sequence of bounded measures on $\mathbb{R}$. Hence by (5.1) we deduce that $a_{mn} = (f_m, \phi(\cdot - n)) \to 0$ and $|a_{mn}| \leq C(n^2 + 1)^p$. From this it follows that if $f_m \to f$ in $\mathcal{S}'_r$, the coefficients of f are of polynomial growth and hence its series $\sum_n a_n \phi(t-n) \in T_0$.

Just as in the L^2 case, we can express each $f_1 \in T_1$ uniquely as $f_1 = f_0 + g_0$ where $f_0 \in T_0$ and $g_0 \in U_0$, in spite of the fact that we do not have orthogonality. Indeed since

$$a_n^0 = (f_1, \phi(t-n)) = \sum_k c_k a_{2n+k}^1 = \sum_j a_j^1 c_{j-2n}, \tag{5.5a}$$

and

$$
\begin{aligned}
b_n^0 = (f_1, \psi(t-n)) &= \sum_k (-1)^k c_{1-k} a_{2n+k}^1 \\
&= \sum_j a_j^1 c_{2n+1-j} (-1)^j,
\end{aligned}
\tag{5.5b}
$$

we have the same decomposition algorithm as in the L^2 case (3.20). Similarly the reconstruction algorithm is given by (3.19) using the same argument.

Thus we use (5.5a) and (5.5b) to define f_0 and g_0 respectively and (3.6) to deduce that $f_1 = f_0 + g_0$. We summarize in the following.

THEOREM 5.1.

Let the scaling function $\phi \in S_r$ *satisfy a dilation equation (3.3) with* $c_k = O(|k|^{-p})$ *for all* $p \in I\!N$; *let* ϕ *have an associated multiresolution analysis in* $L^2(I\!R)$; *let* $\psi \in S_r$ *be the mother wavelet given by (3.16). Then there exists a multiresolution analysis (5.3) of closed dilation subspaces* $\{T_m\}$ *whose union is dense in* S_r'; *the closed subspaces* U_m *(of Definition 5.1) are complementary subspaces of* T_m *in* T_{m+1} *and*

$$T_m = U_0 \oplus U_1 \oplus \cdots \oplus U_m \oplus T_0$$

where $\oplus$ *denotes the nonorthogonal direct sum.*

The only thing we have not shown is that $\bigcup_m T_m$ is dense in S_r'. This follows from the fact that $L^2(I\!R)$ is dense in S_r' and $\bigcup_m V_m \subset \bigcup_m T_m$.

The one property of the usual multiresolution analysis that is lacking is the intersection property $\bigcap_m V_m = \{0\}$. By the moment property of the reproducing kernel $q_m(x,t)$,

$$\int_{-\infty}^{\infty} q_m(x,t) t^k \, dt = x^k, \quad 0 \le k \le r,$$

it follows that any polynomial of degree $\le r$ belongs to each T_m and hence to $\bigcap_m T_m$. Thus, the pure wavelet series of $f \in S_r'$,

$$\sum_{m,n} b_{mn} \psi_{mn}(t),$$

does not necessarily converge to f. However, the mixed expansion,

$$f = \sum_n a_n \phi(\cdot - n) + \sum_{m=0}^{\infty} \sum_n b_{mn} \psi_{mn},$$

does converge to f in the sense of S_r'.

5.2 Wavelets based on distributions

The two methods we have for constructing scaling functions and wavelets, the method based on dilation equations, and the one based on orthogonalization, can be extended to distributions as well. The latter, however, requires a Hilbert space structure, for which we take the Sobolev spaces H^{-s}, $s > \frac{1}{2}$. The dilation equation approach does not, at least in the first steps, and we treat it first.

5.2.1 Distribution solutions of dilation equations

The solution to the dilation equation (3.3)

$$\phi(t) = \sqrt{2} \sum_k c_k \phi(2t - k)$$

may be considered a fixed point of the map

$$(T\phi)(t) = \sqrt{2} \sum_k c_k \phi(2t - k).$$

This in turn may be expressed as

$$T\phi = D_2(k * \phi)$$

where k is the distribution

$$k(t) = \sqrt{2} \sum_k c_k \delta(t - k), \tag{5.6}$$

and D_2 is the dilation operator $D_2\phi = \phi(2\cdot)$.

We assume, as we have in the last section, that $\{c_k\}$ is rapidly decreasing, real, and that

$$\frac{1}{\sqrt{2}} \sum_k c_k = 1. \tag{5.7}$$

T is well defined on $\mathcal{S}'$ and maps $\mathcal{S}'$ into $\mathcal{S}'$. This can be seen by taking the Fourier transform of $k * \phi$, which is

$$(\hat{k} \cdot \hat{\phi}\omega) = \sqrt{2} \sum_k c_k e^{-ik\omega} \hat{\phi}(\omega) = 2m_0(\omega)\hat{\phi}(\omega). \tag{5.8}$$

Since $c_k = O(|k|^{-p})$ for all $p \in \mathbb{N}$, $m_0(\omega) \in C^\infty$ and is periodic and hence is a multiplier in $\mathcal{S}'$. The series in (5.8) converges in $\mathcal{S}'$ and hence its inverse Fourier transform does too.

THEOREM 5.2
Let T be the operator on S' given by $T\phi = D_2(k * \phi)$ where k is as in (5.6) and $\{c_k\}$ is rapidly decreasing and satisfies (5.7). Then there exists a unique $g \in S'$ such that $Tg = g$ where $\hat{g}$ is continuous and of polynomial growth, and $\hat{g}(0) = 1$.

The proof is similar to that in [V] and in [C-H] under different hypotheses. The operator T under the Fourier transform, maps into

$$(\hat{T}\hat{\phi})(\omega) = m_0\left(\frac{\omega}{2}\right)\hat{\phi}\left(\frac{\omega}{2}\right).$$

If we iterate $\hat{T}$ k times, we obtain

$$\hat{T}^k\hat{\phi}(\omega) = \prod_{j=1}^{k} m_0(2^{-j}\omega)\hat{\phi}(2^{-k}\omega) = G_k(\omega)\hat{\phi}(2^{-k}\omega).$$

Since $m_0 \in C^\infty$ and $m_0(0) = 1$ by (5.7), it follows that for $K > 0$, there is a constant C_1 such that

$$|m_0(\omega) - 1| \leq C_1|\omega|, \qquad -K \leq \omega \leq K$$

and hence

$$\left|\sum_{j=m+1}^{n} m_0(2^{-j}\omega) - 1\right| \leq C_1 K 2^{-m} \sum_{j=1}^{n-m} 2^{-j} < C_1 K 2^{-m}$$

for $0 < m < n$, $-K \leq \omega \leq K$. From this we deduce that $\{G_k(\omega)\}$ is uniformly Cauchy on $[-K, K]$ and hence converges uniformly to a continuous function $G(\omega)$, and

$$\hat{T}G(\omega) = G(\omega), \qquad \omega \in \mathbb{R}.$$

In order to find the growth of G, we use the procedure in [V]; we choose $C_2 \geq 1$ such that

$$|G_k(\omega)| \leq C_2 \quad \text{for} \quad |\omega| \leq 1, \quad k \in \mathbb{N},$$

and let $C_3 = \frac{1}{\sqrt{2}}\sum |c_n| \geq 1$. Then $|m_0(\omega)| \leq C_3$, and it follows that for $2^{p-1} \leq |\omega| < 2^p$, $k > p > 0$,

$$|G_k(\omega)| = \left|\prod_{j=1}^{p} m_0(2^{-j}\omega)\right| |G_{k-p}(2^{-p}\omega)| \tag{5.9}$$

$$\leq C_3^p C_2$$

and similarly for $k \leq p$

$$|G_k(\omega)| \leq C_3^k \leq C_3^p C_2.$$

The inequality $2^{p-1} \leq |\omega|$ is now converted to one involving C_3,

$$C_3^{p-1} = 2^{(p-1)M} \leq |\omega|^M$$

where $M = \log_2 C_3 \geq 0$. Thus (5.9) becomes

$$|G_k(\omega)| \leq C_3 C_2 |\omega|^M, \qquad 1 \leq |\omega|.$$

By taking the limit we conclude that $G(\omega)$ is of polynomial growth. Hence $G \in \mathcal{S}'$ and its inverse Fourier transform $g \in \mathcal{S}'$ and is a fixed point of T.

Since T is linear, any multiple of g is also a fixed point, and we can take g such that $\hat{g}(0) = 1$. Then

$$\hat{g}(\omega) = \hat{T}^k \hat{g}(\omega) = G_k(\omega)\hat{g}(2^{-k}\omega) \to G(\omega)\hat{g}(0) = G(\omega)$$

uniformly for $-K \leq \omega \leq K$. This gives us the desired uniqueness. □

COROLLARY 5.3
Let $g(t)$ be the distribution solution to the dilation equation satisfying $\hat{g}(0) = 1$. Then $g \in H^{-s}$ for $s > \log_2 \Sigma |c_n|$.

The proof follows from the fact that

$$|\hat{g}(\omega)| \leq C(1 + |\omega|)^M$$

and hence $g \in H^{-s}$ for $s > M + \frac{1}{2}$.

Although the solution $g \in \mathcal{S}'$, it may not have a "nice" structure. If the dilation equation contains only a finite number of nonzero terms, then g has compact support. If it has only one term, say $c_0 = \sqrt{2}$, then the operator is $(T\phi)(t) = 2\phi(2t)$, and the fixed point may be found explicitly. It is $\phi(t) = \delta(t)$ since $2\delta(2t) = \delta(t)$.

If exactly two terms are nonzero, say c_0 and c_1, then there is never a distribution solution of the form

$$\phi(t) = \sum_{k=0}^{n} a_k \delta(t - t_k).$$

We first observe that the support of ϕ must be in [0,1], and hence we may take $0 \leq t_0 < t_1 < t_2 < \cdots < t_n \leq 1$. From the dilation equation we have

$$\sum_{k=0}^{n} a_k \delta(t - t_k) = \sqrt{2} \sum_{k=0}^{n} \left\{ c_0 \frac{a_k}{2} \delta\left(t - \frac{t_k}{2}\right) + c_1 \frac{a_k}{2} \delta\left(t - \frac{t_k + 1}{2}\right) \right\}.$$

$$(5.10)$$

The coefficients of $\delta(t - t_k)$ on both sides must be equal. If $t_0 = 0$ then $a_0 = \sqrt{2}c_0\frac{a_0}{2}$ and $a_0 = 0$ since $c_0 \neq \sqrt{2}$. If $t_0 > 0$ then $\sqrt{2}\frac{c_0}{2}a_0 = 0$ and $a_0 = 0$ again. Since $\frac{t_k}{2} < t_k < \frac{t_k+1}{2}$, the same argument can be repeated to conclude that $a_1 = a_2 = \cdots = a_n = 0$. We can also extend the argument to more than two non-zero c_k's. In fact we have used only the first sum on the right hand side of (5.10).

PROPOSITION 5.4
The distribution solution to the dilation equation (3.3) has no solution of the form (5.10) if $c_n = 0$ for $|n| \geq I\!N$, but at least two c_n's are not zero.

5.2.2 A partial distributional multiresolution analysis

In Section 5.1, we introduced a multiresolution analysis of S_r' with a ladder of subspaces $\{T_m\}$ each of which was composed of continuous functions of polynomial growth. We now extend this to the distribution solutions of the dilation equation. We define T_0 in a similar way:

$$T_0 = \left\{ f \in \mathcal{S}' \mid f = \sum_n a_n\phi(\cdot - n), \quad a_n \in \ell^2 \right\},$$

where ϕ is the solution given in Theorem 5.2. The series $\sum_n a_n\phi(\cdot - n)$ converges in $\mathcal{S}'$ since its Fourier transform is given by

$$\sum_n a_n e^{-i\omega n}\hat{\phi}(\omega) = \alpha(\omega)\hat{\phi}(\omega),$$

where $\alpha \in L^2(0, 2\pi)$ and is 2π periodic, and hence converges in the sense of $\mathcal{S}'$. Its dilation space $T_1 = \{f \in \mathcal{S}' \mid f(2^{-1}\cdot) \in T_0\}$ contains T_0 since $f \in T_0$ is given by

$$f = \sum_n a_n\phi(\cdot - n) = \sum_n \left(\sum_k a_k c_{n-2k} \right) \sqrt{2}\phi(2\cdot -n)$$

by the now familiar calculation. Since $\{c_k\}$ is rapidly decreasing the inner series converges to a sequence in ℓ^2. Similarly the other dilation spaces T_m are defined and give us the usual ladder of subspaces

$$\cdots \subset T_{-1} \subset T_0 \subset T_1 \subset \cdots \subset \mathcal{S}'.$$

However it is not clear that $\bigcup_m T_m$ is dense in $\mathcal{S}'$ nor can we have the orthogonal complement with a Hilbert space structure. In the next section we consider one case in which such a structure is possible.

5.3 Distributions with point support

In this section we restrict ourselves to scaling functionals in a Sobolev space and introduction a multiresolution analysis of H^{-1} composed of dilation subspaces $\{V_m\}$ where

$$V_0 = \{f \in H^{-1} \mid \operatorname{supp} f \in \mathbb{Z}\}.$$

Then clearly

$$\cdots \subset V_{-1} \subset V_0 \subset V_1 \subset \cdots \subset V_m \subset \cdots H^{-1}.$$

It can also be shown that

$$f(t) = \sum_k a_k \delta(t-k), \qquad \{a_k\} \in \ell^2$$

belongs to V_0, and in fact [W1],[W2] characterizes V_0. This also shows that

$$\overline{\bigcup_m V_m} = H^{-1}$$

since partial sums of dilations of such $f(t)$ are dense in H^{-1}. This comes from the fact that their Fourier transforms are trigonometric polynomials dense in $L^2(\mathbb{R}, (\omega^2 + 1)^{-1})$. However

$$\bigcap_{m=-\infty}^{\infty} V_m = \{C\delta\} \neq \{0\}$$

since $\delta \in V_m$ for all m.

If we try to define the spaces W_m as in the usual approach in Chapter 3, we are confronted with the fact that H^{-s} is not homogeneous; i.e., dilations are not isometries on H^{-s}. However they are homeomorphisms that map H^{-s} into itself with a modified norm. Hence we introduce the inner product on H^{-1} given by

$$\langle f, g \rangle_{-1,\alpha} := \frac{1}{2\pi} \int_{-\infty}^{\infty} \hat{f}(\omega) \overline{\hat{g}(\omega)} \, (\omega^2 + \alpha^2)^{-1} \, d\omega, \qquad \alpha > 0.$$

We note that the V_m are the same whatever α is, but the orthogonal complement of V_0 in V_1 depends on α. We denote by $W_{0,\alpha}$ this orthogonal complement with respect to this inner product with $W_{m,\alpha}$ the corresponding dilation spaces.

We now need to construct an orthonormal basis of V_0 composed of translation of some ϕ. A modification of Meyer's procedure does the trick. Let $\phi \in V_0$; then we find

$$\delta_{0k} = \langle \phi, \phi(\cdot - k)\rangle_{-1,\alpha} = \int_{-\infty}^{\infty} |\hat{\phi}(\omega)|^2 e^{-i\omega k}(\omega^2 + \alpha^2)^{-1}\, d\omega$$

$$= \frac{1}{2\pi}\int_0^{2\pi}\left\{\sum_n |\hat{\phi}(\omega + 2\pi n)|^2 [(\omega + 2\pi n)^2 + \alpha^2]^{-1}\right\} e^{-i\omega k}\, d\omega,$$

and δ_{0k} is the Fourier coefficient of the function in $\{\ \}$, which belongs to $L^2(0, 2\pi)$. Hence it must be equal to its Fourier series (in the L^2 sense),

$$\sum |\hat{\phi}(\omega + 2\pi n)|^2 \left[\sum(\omega + 2\pi n)^2 + \alpha^2\right]^{-1} = \sum_k \delta_{0k}e^{i\omega k} = 1. \quad (5.11)$$

But $|\hat{\phi}|^2$ is periodic (2π) and hence we may solve (5.11) for $|\hat{\phi}|^2$, which is

$$|\hat{\phi}(\omega)|^2 = \left[\sum_n ((\omega + 2\pi n)^2 + \alpha^2)^{-1}\right]^{-1}. \quad (5.12)$$

The positive square root of (5.12) is one solution; the general solution is obtained by multiplying it by a unimodular periodic function. We denote the positive solution by $\hat{\phi}(\omega, \alpha)$,

$$\hat{\phi}(\omega, \alpha) := \left[\sum_n ((\omega + 2\pi n)^2 + \alpha^2)^{-1}\right]^{-1/2} \quad (5.13)$$

and shall concentrate on it. We have

PROPOSITION 5.5
Let $\phi(t, \alpha)$ be given by (5.13); then $\{\phi(t - n, \alpha)\}$ is an orthonormal basis of V_0 with the inner product $\langle\,,\,\rangle_{-1,\alpha}$.

The fact that it is complete in V_0 follows easily from the definition.

The next step is to find an orthonormal basis of $W_{0,\alpha}$, i.e., the wavelets themselves. We look for an element $\psi \in W_{0,\alpha}$ whose translates are orthonormal and complete in $W_{0,\alpha}$. This may be translated into conditions similar to (5.11);

(i) $\quad \sum_n \hat{\psi}(\omega + 2\pi n)((\omega + 2\pi n)^2 + \alpha^{-2})^{-1} = 0.$

(ii) $\quad \sum_n |\hat{\psi}(\omega + 2\pi n)|^2((\omega + 2\pi n)^2 + \alpha^2)^{-1} = 1.$

(5.14)

The first condition is a restatement of the fact that $\psi(t)$ is orthogonal to $\delta(t-k)$ for all $k \in \mathbb{Z}$.

Now however $\hat\psi(\omega)$ is periodic (π) since $\psi \in V_1$; thus both series in (5.14) may be simplified by separating the even and odd terms, which gives us

$$
\begin{aligned}
\text{(i)} \quad & \hat\psi(\omega)a(\omega,\alpha) + \hat\psi(\omega+\pi)a(\omega+2\pi,\alpha) = 0, \\
\text{(ii)} \quad & |\hat\psi(\omega)|^2 a(\omega,\alpha) + |\hat\psi(\omega+\pi)|^2 a(\omega+2\pi,\alpha) = 1
\end{aligned}
\tag{5.15}
$$

where

$$
a(\omega,\alpha) := \sum_n ((\omega+4\pi n)^2 + \alpha^2)^{-1}.
\tag{5.16}
$$

These two equations are easily solved for $|\hat\psi|^2$,

$$
|\hat\psi(\omega)|^2 = \frac{a(\omega+2\pi,\alpha)}{a(\omega,\alpha)(a(\omega+2\pi,\alpha)+a(\omega,\alpha))},
$$

but the positive square root does not satisfy (5.15)(i), but rather must be multiplied by $e^{-i\omega/2}$. We define

$$
\hat\psi(\omega,\alpha) := e^{-i\omega/2}[a(\omega+2\pi,\alpha)/a(\omega,\alpha)(a(\omega+2\pi,\alpha)+a(\omega,\alpha))]^{1/2}.
\tag{5.17}
$$

PROPOSITION 5.6
Let $\psi(t,\alpha)$ be given by (5.17); then $\{\psi(t-n,\alpha)\}$ is an orthonormal basis of $W_{0,\alpha}$.

Again the completeness is straightforward. To find an orthonormal basis of the dilation space V_1, we choose α' such that the map $D_2 f = \sqrt{2}f(2\cdot)$ is a multiple of an isometry from V_0 to V_1. The proper choice is $\alpha' = 2\alpha$, since

$$
\|f\|_{-1,\alpha} = \frac{1}{2}\|D_2 f\|_{-1,2\alpha},
$$

and hence $\{2^{3/2}\phi(2t-n,\alpha'^{1/2})\}$ is an orthonormal basis of V_1.

We now start with $\alpha' = 1$ in each V_m and find the corresponding α in V_0 so that we can get an orthonormal basis of V_m. This gives us

$$
\phi_{m,n}(t) = 2^{3m/2}\phi(2^m t - n, 2^{-m}), \qquad n = 0, \pm 1, \cdots.
\tag{5.18}
$$

For the orthogonal complement we begin with $W_0 = W_{0,1}, \cdots, W_m = W_{m,1}, \cdots$ and thereby obtain similarly

$$
\psi_{m,n}(t) = 2^{3m/2}\psi(2^m t - n, 2^{-m}), \qquad n = 0, \pm 1, \cdots.
\tag{5.19}
$$

COROLLARY 5.7
Let $\phi_{m,n}(t)$ and $\psi_{m,n}(t)$ be given by (5.18) and (5.19) respectively; then $\{\phi_{0,n}(t)\}_{n=-\infty}^{\infty}$ and $\{\psi_{m,n}(t)\}_{m=0,\,n=-\infty}^{\infty}$ form an orthonormal basis of H^{-1} with the usual inner product.

Hence we have the usual multiresolution analysis

$$H^{-1} = V_0 \oplus \bigoplus_{m=0}^{\infty} W_m = \{\delta\} \oplus \bigoplus_{m=-\infty}^{\infty} W_m \qquad (5.20)$$

where $\{\delta\}$ is the space spanned by the δ function.

Our theory is almost complete. We still need to get simpler expressions for $\psi(t,\alpha)$ by summing the series for $a(\omega,\alpha)$ of (5.16). We begin with the fact that

$$\sum_{n=-\infty}^{\infty} \frac{1}{x+2\pi n} = \frac{1}{2}\frac{\cos\frac{x}{2}}{\sin\frac{x}{2}}$$

from the Fourier series of e^{-ixt} on $(0,1)$ evaluated at $t=0$. Then

$$
\begin{aligned}
a(\omega,\alpha) &= \frac{i}{4\alpha}\sum_{n}\frac{1}{\frac{\omega+\alpha i}{2}+2\pi n} - \sum\frac{1}{\frac{\omega-\alpha i}{2}+2\pi n} \\
&= \frac{i}{8\alpha}\left\{\frac{\cos(\omega+i\alpha)/4}{\sin(\omega+i\alpha)/4} - \frac{\cos(\omega-\alpha i)/4}{\sin(\omega-\alpha i)/4}\right\} \\
&= \frac{i}{8\alpha}\frac{\sin(-i\alpha/2)}{\sin^2\omega/4\cos^2\alpha i/4 - \cos^2\omega/4\sin^2\alpha i/4} \\
&= \frac{1}{4\alpha}\frac{\sinh\alpha/2}{\cos\alpha/2-\cos\omega/2}.
\end{aligned}
$$

This is substituted into (5.17) to obtain

$$\hat{\psi}(\omega,\alpha) = e^{-i\omega/2}(\cosh\alpha/2 - \cos\omega/2)\sqrt{\frac{4\alpha}{\sinh\alpha}} \qquad (5.21)$$

after a little manipulation. The inverse Fourier transform is easily found to be, for $\alpha = 2^{-m}$

$$\psi(t,2^{-m}) = C\left\{\cosh(2^{-m-1})\delta(t-\tfrac{1}{2}) - \tfrac{1}{2}\delta(t) - \tfrac{1}{2}\delta(t-1)\right\} \qquad (5.22)$$

where C is a positive constant.

We can also find $\phi(t,\alpha)$ to be given by

$$\hat{\phi}(\omega,\alpha) = \left[\frac{2\alpha(\cosh\alpha - \cos\omega)}{\sinh\alpha}\right]^{1/2}.$$

However this leads to an infinite Fourier series whose sum is not as simple [W1]. This really doesn't matter that much since the important result is (5.22), which gives an orthonormal basis of H^{-1} each of whose elements has support on three points (or fewer in the case of $\delta(t)$).

5.4 Problems

1. Find the scaling function expansion $\sum_n a_n \phi(t-n)$ of $\delta(t-\alpha)$ for any 1-regular ϕ.

2. Repeat Problem 1 for $t^k \delta(t-\alpha)$ $k \le r$ for any r-regular ϕ.

3. Use the result in Problem 2 to show that

$$(t^k - \alpha^k)g_m(x,t) \to 0 \qquad \text{as } m \to \infty$$

   weakly.

4. For 1-regular ϕ, the function $x \in T_0$. (Why?) Show that its expansion is

$$x = \sum (n+\mu)\phi(x-n),$$

   where $\mu = \int\limits_{-\infty}^{\infty} t\phi(t)dt$.

5. Conjecture and prove a similar result to Problem 4 for x^k, $k \le r$ when ϕ is r-regular.

6. Find the approximation to $\delta(t-\alpha)$ in terms of its expansion in

$$T_0 \oplus U_0 \oplus U_1.$$

7. Let the coefficients in the dilation equation be $c_0 = c_1 = c_2 = c_3 = \frac{1}{2\sqrt{2}}$. Find a closed form expression for $m_0(\omega)$. Then use the argument of Theorem 5.2 to show that $\hat{\phi}(\omega)$ is bounded.

8. It can be shown [C-H] that $\phi(t) \in L^2(\mathbb{R})$ for ϕ of Problem 7. Show that $\{\phi(t-n)\}$ is not orthogonal however.

9. Find the expansion coefficients of $f(x) = e^{-|x|}$ with respect to the wavelet basis $\{\psi_{m,n}\}$ of Corollary 5.7 in H^{-1}. (Caution: this uses the Sobolev inner product.)

6

Orthogonal Polynomials

Just as with trigonometric series, there is a large body of literature on orthogonal polynomials. The most complete source is Szegö's book [Sz], which in particular gives a detailed exposition of Jacobi polynomials (see also [C], [As], and [J]). In this chapter we present elements of the theory concentrating on those parts that are useful in applications or which are related to topics covered in other chapters. We first present some of the general theory including interpolating polynomials and then take up the classical Jacobi, Laguerre, and Hermite systems.

6.1 General theory

We begin with a nonnegative (Borel) measure μ on the interval (a, b) such that $\mu(a, b) < \infty$. The space in which we now work is $L^2(\mu; (a, b))$ consisting of all functions such that

$$\int_a^b |f|^2 d\mu < \infty.$$

The notation $\langle f, g \rangle_\mu$ will now mean the inner product in this space with $\|f\|_\mu$ denoting the norm. In most cases $d\mu$ will be given by a weight function

$$d\mu(x) = w(x)dx,$$

although in some discrete cases we assume it to be

$$d\mu(x) = \sum a_i \delta(x - x_i)dx.$$

In order that polynomials belong to the space, we shall also require that the moments $\int_a^b x^n d\mu$, $n = 0, 1, \cdots$, exist. This is no problem if (a, b) is a finite interval but it may be if it is infinite.

There generally exist many orthogonal polynomial systems associated with μ, but we shall usually consider systems where $P_n(x)$ is of degree n. This may be ensured by orthogonalizing the sequence

$$1, x, x^2, \cdots, x^n, \cdots.$$

That is, we let $P_0(x) = 1$ and then subtract from x its projection on the space spanned by $P_0(x)$. This gives us $P_1(x)$. We then subtract from x^2 its projection on the space spanned by $P_0(x)$ and $P_1(x)$ to get $P_2(x)$, etc. This will give us a sequence $\{P_n\}$ of polynomials that are of exact degree n and orthogonal with respect to $d\mu$. They may be made into an orthonormal system by normalization

$$p_n(x) = P_n(x)/\|P_n\|_\mu.$$

If the interval (a, b) is finite, and the support of μ is an infinite set, the orthonormal system is always complete. This follows from the Weierstrass approximation theorem [Sz, p.5], which says that a continuous function on $[a, b]$ can be uniformly approximated by polynomials. Since each $f \in L^2(\mu; (a, b))$ can be approximated by continuous functions in the sense of the norm $\| \ \|_\mu$, it follows that f can be approximated by polynomials in the sense of $\| \ \|_\mu$. Furthermore, since the partial sums of the orthogonal polynomial expansion gives the best approximation (Chapter 1) to f, it follows that

$$\left\| f - \sum_{n=0}^{N} \langle f, p_n \rangle_\mu p_n \right\|_\mu \to 0$$

as $N \to \infty$ and hence $\{p_n\}$ is complete.

If the support of μ is a finite set, as happens in interpolation problems, then $1, x, \cdots, x^n, \cdots$ spans a finite dimensional space, and the sequence of orthogonal polynomials terminates after a finite number.

Some examples of orthogonal polynomials (which are usually not normalized) are:

1. The Legendre polynomials $\{P_n(x)\}_{n=0}^{\infty}$, $a = -1$, $b = 1$ and $d\mu(x) = dx$.
2. The Hermite polynomials $\{H_n(x)\}_{n=0}^{\infty}$, $a = -\infty$, $b = \infty$ and $d\mu(x) = e^{-x^2} dx$.
3. The Krawtchouk polynomials $\{K_n(x))\}_{n=0}^{N}$, $a = 0$, $b = N$ and $d\mu(x) = \sum_{k=0}^{N} \binom{N}{k} q^k (1-q)^{N-k} \delta(x-k)$, $0 < q < 1$. Here the closed interval $[O, N]$ is needed since the endpoints have positive mass.

A surprising property of orthogonal polynomials is that they always satisfy a three term <u>recurrence relation</u>

$$xp_n(x) = a_n p_{n+1}(x) + b_n p_n(x) + c_n p_{n-1}(x), \ n = 1, 2, \cdots. \qquad (6.1)$$

To prove this, we first choose a_n such that $x p_n - a_n p_{n+1}$ is a polynomial of degree n, which we may then write as a linear combination of $p_0, p_1, \cdots, p_n$. But $\int_a^b p_k(x) p_n(x) d\mu = 0$ for $k < n - 1$ since p_n is orthogonal to all polynomials of lower degree (in particular to $x p_k(x)$), and the linear combination consists of the terms $b_n p_n(x) + c_n p_{n-1}(x)$.

This is used to find the kernel $K_n(x, y)$ of convergence, which is analogous to the Dirichlet kernel of Fourier series. The partial sums of the expansion of $f \in L^2(\mu; (a, b))$ are

$$
\begin{aligned}
S_n(x) &= \sum_{k=0}^{n} \int_a^b f(y) p_k(y) d\mu(y) p_k(x) \\
&= \int_a^b f(y) \sum_{k=0}^{n} p_k(y) p_k(x) d\mu(y) \qquad (6.2) \\
&= \int_0^b f(y) K_n(x, y) d\mu(y)
\end{aligned}
$$

and the kernel is

$$
\begin{aligned}
(x - y) K_n(x, y) &= \sum_{k=0}^{n} x p_k(x) p_k(y) - \sum_{k=0}^{n} y p_k(y) p_k(x) \\
&= \sum_{k=0}^{n} (a_k p_{k+1}(x) + b_k p_k(x) + c_k p_{k-1}(x)) p_k(y) \\
&\quad - \sum_{k=0}^{n} (a_k p_{k+1}(y) + b_k p_k(y) + c_k p_{k-1}(y)) p_k(x) \\
&= \sum_{k=0}^{n} a_k (p_{k+1}(x) p_k(y) - p_{k+1}(y) p_k(x)) \\
&\quad + \sum_{k=0}^{n} c_k (p_{k-1}(x) p_k(y) - p_{k-1}(y) p_k(x))
\end{aligned}
$$

$$(6.3)$$

where we have taken $c_0 = 0$. We now use the fact that

$$
\int_a^b x p_n p_{n+1} d\mu = \int (a_n p_{n+1}^2 + b_n p_n p_{n+1} + c_n p_{n-1} p_{n+1}) d\mu = a_n
$$

and similarly

$$
c_n = \int_a^b x p_n p_{n-1} d\mu = a_{n-1}.
$$

Thus the sums in (6.3) cancel each other except for one term, and we get

$$
K_n(x, y) = \frac{a_n (p_{n+1}(x) p_n(y) - p_n(x) p_{n+1}(y))}{x - y}, \qquad (6.4)
$$

the Christoffel-Darboux formula.

Just as with trigonometric Fourier series, we can prove pointwise convergence by using $K_n(x,y)$. For example, if $f(x) \in L^2(\mu;(a,b))$ and satisfies a Lipschitz condition of order α at $x_0 \in (a,b)$, $\frac{1}{2} < \alpha \le 1$, $b - a < \infty$, then in many cases

$$f_n(x) := \int_a^b K_n(x_0,y)f(y)d\mu(y) \to f(x_0).$$

The argument is the same as in Chapter 4 for Fourier series and needs only a suitable bound on $p_n(x)$.

This kernel can also be used to determine the behavior of the real zeros of $\{p_n(x)\}$ when the support of μ is an infinite set [Sz]. By taking $y = x$ in (6.4) we obtain

$$\begin{aligned}
K_n(x,x) &= a_n(p'_{n+1}(x)p_n(x) - p'_n(x)p_{n+1}(x)) \\
&= \sum_{k=0}^n p_k^2(x) > 0.
\end{aligned} \quad (6.5)$$

Since by the construction the leading coefficients of $p_n(x)$ are positive and since

$$xp_n(x) - a_np_{n+1}(x)$$

is a polynomial of degree $\le n$, it follows that a_n must be a quotient of two leading coefficients and hence must be > 0. Hence we have

$$p'_{n+1}(x)p_n(x) - p'_n(x)p_{n+1}(x) > 0,$$

from which we deduce that if $p_n(x_0) = 0$,

$$p_{n+1}(x_0)p'_n(x_0) < 0 \quad \text{and} \quad p_{n-1}(x_0)p'_n(x_0) > 0. \quad (6.6)$$

Also from the recurrence formula (6.1) we have

$$a_np_{n+1}(x_0) + \alpha_{n-1}p_{n-1}(x_0) = 0$$

and hence

$$a_np_{n+1}(x_0)p_{n-1}(x_0) = -a_{n-1}p_{n-1}^2(x_0) < 0. \quad (6.7)$$

From (6.6), it follows that the zeros are simple, and p_n and p_{n+1} cannot have common zeros. Let x_0 and x_1 be two successive zeros of $p_n(x)$; then $p'_n(x_0)$ and $p'_n(x_1)$ have opposite signs and by (6.6) so do $p_{n+1}(x_0)$ and $p_{n+1}(x_1)$.

Hence, there is an odd number of zeros of p_{n+1} between x_0 and x_1. By a similar argument involving the other inequality in (6.6), there is an odd number of zeros of $p_n(x)$ between two successive zeros of $p_{n+1}(x)$. Thus, there is exactly one zero between x_0 and x_1. For the largest zero of $p_n(x)$, say x_m, $p_n'(x_m) > 0$ and $p_{n+1}(x_m) < 0$ by (6.6). Since the leading coefficient of p_{n+1} is positive, $p_{n+1}(x) > 0$ for x sufficiently large and hence p_{n+1} must have a zero larger than x_m; similarly it has a zero smaller than the smallest zero of p_m. This gives p_{n+1} one more zero than p_n. Since p_1 has one zero (otherwise $\langle p_1, p_0 \rangle_\mu > 0$), p_2 must have two zeros, p_3, three, ...

PROPOSITION 6.1
The zeros of $p_n(x)$ are real, simple, and in (a,b) and are interlaced with those of $p_{n+1}(x)$.

We have not shown that they lie in (a,b). If only $x_0, x_1, \ldots, x_m$, $m+1 < n$ lie in (a,b) then

$$\int_a^b p_n(x) \prod_{j=0}^m (x - x_j) d\mu(x) \neq 0$$

since the integrand does not change sign in (a,b). But $\prod_{j=0}^m (x - x_j)$ is a polynomial of degree $m < n$ and therefore orthogonal to p_n which gives a contradiction.

If the measure μ is discrete and finite, the orthogonal polynomials can be taken to be interpolation polynomials. That is, if

$$d\mu(x) = \sum_{i=0}^n a_i \delta(x - x_i) dx, \quad a < x_0 < \cdots < x_n < b,$$

then the polynomials $\ell_i(x)$ of degree n, $i = 0, 1, \cdots, n$ where $\ell_i(x_j) = \delta_{ij}$ will be orthogonal with respect to $d\mu$. Indeed, we see that

$$\int_a^b \ell_i(x)\ell_j(x)d\mu(x) = \sum_k a_k \ell_i(x_k)\ell_j(x_k) = a_i \delta_{ij}, \quad i, j = 0, 1, \cdots, n.$$

It is easy to define these polynomials by

$$\ell_i(x) := \prod_{j=0}^n (x - x_j) / \prod_{i \neq j} (x_i - x_j).$$

These $\{\ell_i\}$ constitute a complete orthogonal set in $L^2(d\mu, (a,b))$; they are usually referred to as <u>Lagrange interpolation polynomials</u> at the nodes $x_0, x_1, \cdots, x_n$. Any polynomial of degree n has a representation

$$P(x) = \sum_{i=0}^{n} P(x_i)\ell_i(x).$$

This is a simple example of a sampling theorem that will be taken up in Chapter 9.

6.2 Classical orthogonal polynomials

These classical orthogonal polynomials are usually taken to be on the interval $(-1, 1)$, $(0, \infty)$, or $(-\infty, \infty)$. In the first case, the measure is

$$d\mu = w(x)dx = (1 - x)^\alpha (1 + x)^\beta dx$$

which gives for general $\alpha, \beta > -1$, the Jacobi polynomials, for $\alpha = \beta = 0$, the Legendre polynomials, for $\alpha = \beta = \lambda - \frac{1}{2}$ the Gegenbauer polynomials, for $\alpha = \beta = \frac{1}{2}$, the Tchebychev polynomials. We shall discuss only the Legendre case in detail. On the infinite intervals, we have the Laguerre and Hermite polynomials. More details may be found in [Sz].

6.2.1 Legendre polynomials

The Legendre polynomials $\{P_n\}$ are usually normalized so that $P_n(1) = 1$. They are obtained as before by orthogonalizing $1, x, x^2, \cdots$ with respect to dx. It is easy to find that

$$P_0(x) = 1, \ P_1(x) = x, \ P_2(x) = \frac{1}{2}(3x^2 - 1),$$

but then it becomes a little harder. The others are given most easily by the recurrence formula (6.2), which in this case becomes

$$xP_n(x) = \frac{n+1}{2n+1}P_{n+1}(x) + \frac{n}{2n+1}P_{n-1}(x). \tag{6.8}$$

This does not have the exact form of (6.1) since the $P_n(x)$ do not have norm 1; in fact $\|P_n\|^2 = 1/\left(n + \frac{1}{2}\right)$.

Another approach may be based on the differential equation [Jc, p.49]

$$((1 - x^2)y')' + n(n+1)y = 0 \tag{6.9}$$

which is satisfied for $y = P_n(x)$. Both the recurrence formula and the differential equation may be based on the generating function [Jc, p.45]

$$H(x,r) = (1 - 2xr + r^2)^{-1/2}.$$

The power series expansion of $H(x,r)$ is for $|r| < 1$ given by

$$H(x,r) = \sum_{n=0}^{\infty} P_n(x)r^n. \tag{6.10}$$

The Christoffel-Darboux formula (6.4) becomes

$$K_n(x,y) = \frac{n+1}{2} \frac{P_{n+1}(y)P_n(x) - P_n(y)P_{n+1}(x)}{y-x}.$$

This may be used to obtain a convergence theorem similar to that for trigonometric Fourier series.

PROPOSITION 6.2
Let $f \in L^2(-1,1)$ and satisfy a Lipschitz condition of order $\alpha > \frac{1}{2}$ at $x \in (-1,1)$, then the Legendre expansion

$$\sum_{n=0}^{\infty} a_n P_n(x), \quad a_n = \left(n + \frac{1}{2}\right) \int_{-1}^{1} f(x)P_n(x)dx$$

converges to $f(x)$.

PROOF The proof uses the fact that

$$S_n(x) - f(x) = \int_{-1}^{1} (f(t) - f(x))K_n(x,t)dt$$

and is similar to the one for Fourier series. The necessary bounds on the kernel are obtained from the integral formula for the P_n [Jc, p.61], which is proved by showing that it satisfies the same recurrence formula as the $P_n(x)$,

$$P_n(\cos\theta) = \frac{1}{\pi} \int_0^{\pi} [\cos\theta + i\sin\theta\sin\phi]^n d\phi. \tag{6.11}$$

It is clear from this integral formula that

$$|P_n(\cos\theta)| \leq 1,$$

but it can also be used to prove a stronger inequality

$$|P_n(x)| \leq C_\epsilon n^{-1/2}, \quad n = 1, 2, \cdots, |x|^2 < 1 - \epsilon.$$

This follows from

$$|P_n(\cos\theta)| \leq \frac{1}{\pi}\int_0^\pi \left[\cos^2\theta + \sin^2\theta\cos^2\phi\right]^{n/2} d\phi$$

$$= \frac{2}{\pi}\int_0^{\pi/2}\left[1 - \sin^2\theta\sin^2\phi\right]^{n/2} d\phi$$

$$\leq \frac{2}{\pi}\int_0^{\pi/2}\left[1 - \sin^2\theta\left(\frac{2\phi}{\pi}\right)^2\right]^{n/2} d\phi$$

$$\leq \frac{2}{\pi}\int_0^{\pi/2}\left[e^{-\sin^2\theta(2\phi/\pi)^2}\right]^{n/2} d\phi$$

$$= \frac{2}{\pi}\frac{\pi}{\sin\theta\, 2\sqrt{\frac{n}{2}}}\int_0^{\sin\theta\sqrt{n/2}} e^{-t^2} dt$$

$$\leq \frac{\sqrt{2}}{\sqrt{n}\sin\theta}\int_0^\infty e^{-t^2} dt, \qquad 0 < \theta < \pi.$$

If $|x|^2 = \cos^2\theta < 1 - \epsilon$, then $\sin^2\theta > \epsilon$, and the inequality follows. We then return to the difference, and express it as

$$|S_n(x) - f(x)| \leq \left|\int_{-1}^{x-\delta}\right| + \left|\int_{x-\delta}^{x+\delta}\right| + \left|\int_{x+\delta}^1 (f(t) - f(x))K_n(x,t)dt\right|$$

where δ is sufficiently small that $(x - \delta, x + \delta) \subset (-1, 1)$. The middle integral is dominated by

$$\frac{n+1}{2}\int_{x-\delta}^{x+\delta}|x - t|^\alpha\left|\frac{P_n(x)P_{n+1}(t) - P_n(t)P_{n+1}(x)}{x - t}\right| dt$$

$$\leq C\int_{x-\delta}^{x+\delta}|x - t|^{\alpha-1}dt = \frac{C_2\delta^\alpha}{\alpha},$$

since both x and t are in an interval $(-1 + \epsilon, 1 - \epsilon)$. The other integrals are made up of terms of the form

$$(n+1)\int_{-1}^1 g(x,t)P_n(x)P_{n-1}(t)dt \qquad (6.12)$$

where $g(x, \cdot) \in L^2(-1, 1)$. Since $\left\{\sqrt{n + \frac{3}{2}}P_{n+1}(t)\right\}$ is orthonormal, it follows by Bessel's inequality that

$$\sqrt{n+1}\int_{-1}^1 g(x,t)P_{n+1}(t)dt \to 0 \text{ as } n \to \infty.$$

Also since $\sqrt{n+1}\,P_n(x)$ is bounded in n, the integral in (6.12) is the product of a bounded function and one that converges to 0 and hence also converges to 0 as $n \to \infty$. Hence for any $\epsilon > 0$, we may choose δ so that $\frac{2C}{\alpha}\delta^\alpha < \frac{\epsilon}{2}$ and then choose n_0 so large that the remaining integrals are less than $\frac{\epsilon}{2}$ for $n \geq n_0$, thereby concluding that

$$|S_n(x) - f(x)| < \epsilon \qquad \text{for } n \geq n_0. \qquad \square$$

This same formula (6.11) can be used to find the derivatives of P_n to deduce that

$$\left| \frac{d^k}{d\alpha^k} P_n(\cos\alpha) \right| \leq C\left(n + \frac{1}{2}\right)^k, \qquad n = 0, 1, 2, \cdots \qquad (6.13)$$

for some constant C. This gives results similar to Proposition 6.2 for distribution expansions.

PROPOSITION 6.3
Let $f \in S'$ and have support in the interior of $(-1, 1)$, let $a_n = (f, P_n\theta)$, where $\theta \in S$, $\theta(x) = 1$ for $x \in \text{supp}(f)$, and $\text{supp}(\theta) \subset (-1, 1)$. Then
  (i) $a_n = O(n^p)$ for some $p \in \mathbb{Z}$
  (ii) $\sum_n a_n P_n\theta$ converges to f in S'.

PROOF If f has support on $[\alpha, \beta]$, then there exists a function $G \in L^2(-1, 1)$ with the same support and $f = D^p G + \sum_{i=0}^{p-1} C_i \delta^{(i)}$ (see Chapter 2). We then use integration by parts (i.e., the definition of distribution derivative) on the coefficients

$$a_n = (D^p G, P_n\theta) + \sum_{i=0}^{p-1} C_i P_n^{(i)}(0)$$

$$= (-1)^p (G, P_n^{(p)}\theta) + \sum_{i=0}^{p-1} C_i P_n^{(i)}(0)$$

which by (6.13) satisfy the conditions needed for (i).
  For (ii) we have to show that

$$\sum_{n=0}^{\infty} a_n (P_n\theta, \phi)$$

converges for each $\phi \in S$. Since

$$n(n+1)(P_n, \theta\phi) = -(((1-x^2)P_n')', \theta\phi) = -(P_n, ((1-x^2)(\theta\phi)')')$$

and may be iterated any number of times, $(P_n, \theta\phi) = 0(n^{-q})$ for any $q \in \mathbb{N}$. Thus the series converges. That it converges to f involves the same sort of considerations and Parseval's equality, $(f, \theta\phi) = (-1)^p(G, (\theta\phi_1^{(p)}) + \sum C_i(\theta\phi)^{(i)}(0)$. $\square$

6.2.2 Jacobi polynomials

The Jacobi polynomials $\{P_n^{(\alpha,\beta)}\}$ are orthogonal with respect to the weight function $w(x) = (1-x)^\alpha(1+x)^\beta$, $-1 < \alpha, \beta$. As with the Legendre polynomials, they may be obtained by orthogonalizing $\{1, x, x^2, \cdots\}$. This gives us $P_0^{(\alpha,\beta)}(x) = 1$ while

$$P_1^{(\alpha,\beta)}(x) = \frac{\alpha + \beta + 2}{2}x + \frac{\alpha - \beta}{2}$$

with the usual normalization given by $P_n^{(\alpha,\beta)}(1) = \binom{n+\alpha}{n}$.

The remaining $P_n^{(\alpha,\beta)}$ may be obtained from the recurrence formula

$$xP_n^{(\alpha,\beta)} = a_n P_{n+1}^{(\alpha,\beta)} + b_n P_n^{(\alpha,\beta)} + c_n P_{n-1}^{(\alpha,\beta)}, \tag{6.14}$$

where

$$a_n = 2(n + 1)(\lambda + n + 1)/(\lambda + 2n + 1)(\lambda + 2n + 2),$$
$$b_n = (\beta^2 - \alpha^2)/(n + 2n)(\lambda + 2n + 2),$$
$$c_n = 2(\alpha + n)(\beta + n)/(\lambda + 2n)(\lambda + 2n + 1),$$

and $\lambda = \alpha + \beta$. The differential equation is

$$((1 - x^2)y')' + [\beta(1 - x) - \alpha(1 + x)]y' + n(n + \alpha + \beta + 1)y = 0,$$

and the normalization constant $\delta_n^{(\alpha,\beta)} = \left\| P_n^{(\alpha,\beta)} \right\|_w^2$ is

$$\delta_n^{(\alpha,\beta)} = \frac{2^{\lambda+1}}{\lambda + 2n + 1} \frac{\Gamma(\alpha + n + 1)\Gamma(\beta + n + 1)}{n!\Gamma(\lambda + n + 1)}.$$

Other formulae for the Legendre polynomials have analogues for the Jacobi polynomials. Convergence of the expansion of functions and distributions is also proved in the same way. An exhaustive treatment of Jacobi polynomials may be found in [Sz].

6.2.3 Laguerre polynomials

The Laguerre polynomials $\left\{ L_n^{(\alpha)}(x) \right\}$ are orthogonal on $(0, \infty)$ with respect to the weight function $w(x) = x^\alpha e^{-x}$, $\alpha > -1$. Their first two elements are

$$L_0^{(\alpha)}(x) = 1, \quad L_1^{(\alpha)}(x) = -x + \alpha + 1,$$

while the recurrence formula is

$$x L_n^{(\alpha)} = -(n+1)L_{n+1}^{(\alpha)} + (2n+1+\alpha)L_n^{(\alpha)} - (n+\alpha)L_{n-1}^{(\alpha)}. \tag{6.15}$$

They satisfy the differential equation

$$(xy')' + (\alpha - x)y' + ny = 0,$$

and their normalization factor $\delta_n^\alpha = \int_0^\infty \left| L_n^\alpha(x) \right|^2 w(x)dx$ is

$$\delta_n^\alpha = \Gamma(n+\alpha+1)/\Gamma(n+1).$$

Since the interval is infinite, they are not automatically a complete system. However, in this case the completeness may be proved using the Laplace transform:

$$F(s) = \int_0^\infty e^{-st} f(t)dt. \tag{6.16}$$

If $f(x)(x^2+1)^{-p} \in L^2(0,\infty)$ for some $p \in \mathbb{Z}$, then $F(s)$ is a holomorphic on the half-plane Re $s > 0$. We have to show that if

$$\int_0^\infty e^{-x/2} x^n x^{\alpha/2} g(x)dx = 0, \quad n = 0, 1, 2, \cdots, \tag{6.17}$$

for $g \in L^2(0,\infty)$, then $g = 0$ a.e. This would imply that $\left\{ x^n x^{\alpha/2} e^{-x/2} \right\}$ and hence $\left\{ L_n^{(\alpha)}(x) x^{\alpha/2} e^{-x/2} \right\}$ are complete in $L^2(0,\infty)$.

We now take $f(t) = t^{\alpha/2} g(t)$ in the Laplace transform (6.16) and use the fact that

$$F'(s) = -\int_0^\infty t e^{-st} f(t)dt$$

to express (6.17) as

$$(-1)^n F^{(n)} \left(\frac{1}{2} \right) = 0, \quad n = 0, 1, \cdots.$$

This implies that $F(s) \equiv 0$ for Re $s > 0$ and hence that $g(t) = 0$, a.e.

Again local and global convergence theorems can be proved by the same tricks. If anything the formulae and bounds are simpler than in the Jacobi case and may again be found in [Sz].

6.2.4 Hermite polynomials

The Hermite polynomials $\{H_n(x)\}$ are orthogonal on $(-\infty, \infty)$ with respect to the weight function $w(x) = e^{-x^2}$. They may be obtained by orthogonalizing $1, 2x, (2x)^2, \cdots$ with respect to this weight, or as before, using a recurrence formula together with

$$H_0(x) = 1, \quad H_1(x) = 2x.$$

It is

$$x\, H_n(x) = \frac{1}{2} H_{n+1}(x) + n H_{n-1}(x), \quad n = 1, 2, \cdots.$$

An alternate approach uses the differential equation

$$y'' - 2xy' + 2ny = 0$$

whose only polynomial solution is a multiple of $H_n(x)$. In the usual normalization, $H_n(x)$ has leading coefficient 2^n; this makes its norm equal to

$$\int_{-\infty}^{\infty} H_n^2(x) e^{-x^2} dx = \pi^{1/2} 2^n n!, \quad n = 0, 1, 2, \cdots.$$

Rather than use these polynomials for our analysis we combine them with the weight to get the <u>Hermite functions</u>

$$h_n(x) = \frac{H_n(x) e^{-x^2/2}}{\pi^{1/4} 2^{n/2} (n!)^{1/2}}, \quad n = 0, 1, 2, \cdots, x \in \mathbb{R}.$$

These form a complete orthonormal sequence in $L^2(\mathbb{R})$ and are widely known as forming a solution to the linear oscillator equation,

$$(x^2 - D^2)y = (2n + 1)y, \quad n = 0, 1, \cdots.$$

However, they also have a number of other uses in mathematics that concern us here. As we observed in Chapter 2, they belong to the space S and can be used to characterize the space of tempered distributions S'.

The recurrence formula for the polynomials can be restated in terms of the $\{h_n(x)\}$ as

$$x\, h_n(x) = \sqrt{\frac{n}{2}}\, h_{n-1}(x) + \sqrt{\frac{n+1}{2}}\, h_{n+1}(x). \qquad (6.17a)$$

In addition the derivatives satisfy

$$h'_n(x) = \sqrt{\frac{n}{2}}\, h_{n-1}(x) - \sqrt{\frac{n+1}{2}}\, h_{n+1}(x).\qquad(6.17b)$$

The $h_n(x)$ are uniformly bounded for real x. Indeed, we have

$$|h_n(x)| \le Cn^{-1/12}, \qquad n = 1, 2, \cdots$$

for a constant C ([Sz, p.242]). With this bound and the recurrence formulae we can get a pointwise convergence theorem.

PROPOSITION 6.4
Let $f(x) \in L^2(\mathbb{R})$, *be absolutely continuous, and satisfy* $(x + D)f(x) \in L^2(\mathbb{R})$; *then the Hermite series* $\sum a_n h_n(x)$ *of* $f(x)$ *converges uniformly on* $\mathbb{R}$ *to* $f(x)$.

PROOF   By taking the difference between (6.17a) and (6.17b) we obtain

$$(x - D)h_n(x) = \sqrt{2n + 2}\, h_{n+1}(x)\qquad(6.18)$$

and hence, the coefficients of $(x + D)f$ are $a_{n+1}\sqrt{2n + 2}$.
   Then the Hermite series of $f(x)$ can be bounded by

$$\sum_{n=1}^{\infty} |a_n h_n(x)| \le \sum_n \left| a_n \sqrt{2n}\, \frac{h_n(x)}{\sqrt{2n}} \right|$$

$$\le \left\{ \sum_n \left| a_n \sqrt{2n} \right|^2 \right\}^{1/2} \left\{ \sum_n \frac{h_n^2(x)}{2n} \right\}^{1/2}$$

$$\le \|(x + D)f\| \left\{ \sum_n \frac{C}{(2n)^{7/6}} \right\}^{1/2},$$

and hence the series converges absolutely and uniformly on $\mathbb{R}$. The limit function, say $g(x)$, must be bounded since the partial sums are bounded.
   The Hermite coefficients of g must be given by

$$\langle g, h_m \rangle = \sum_n a_n \langle h_n, h_m \rangle = a_m,$$

i.e., must be the same as those of f. We deduce that $f(x) = g(x)$ by the following uniqueness theorem.

LEMMA 6.5
Let $F(x)$ *be a continuous function of polynomial growth whose Hermite coefficients* $\langle F, h_n \rangle = 0$, $n = 0, 1, \cdots$; *then* $F \equiv 0$.

PROOF Since F is of polynomial growth $\left|F(x)e^{-x^2/2}\right| \le Ce^{-\epsilon|x|}$ for some $\epsilon > 0$ and constant C. Thus, the Fourier transform of $F(x)e^{-x^2/2}$ is analytic in a strip $-\delta < \operatorname{Im} w < \delta$. By the hypothesis we have

$$\int_{-\infty}^{\infty} F(x)e^{-x^2/2}H_n(x)dx = 0, \qquad n = 0, 1, 2, \cdots$$

and, hence, since x^n is a linear combination of $H_0(x), \cdots, H_n(x)$,

$$\int_{-\infty}^{\infty} F(x)e^{-x^2/2}x^n dx = 0, \qquad n = 0, 1, 2, \cdots.$$

Therefore the Fourier transform of $F(x)e^{-x^2/2}$ will have all of its derivatives zero at 0, and hence it must be identically zero. So must $F(x)e^{-x^2/2}$ by the Fourier integral theorem. □ □

This uniqueness theorem may be extended to all of S' by an argument of J. Korevaar [K1].

COROLLARY 6.6
Let $f \in S'$ have Hermite coefficients $(f, h_n) = 0$, $n = 0, 1, \cdots$, then $f \equiv 0$.

PROOF Each $f \in S'$ may be expressed as $f = D^p F$ where F is continuous of polynomial growth. We proceed by induction on p. The result is true for $p = 0$ by the Lemma and we suppose it true for $p - 1$. Then $f = Dg = D(D^{p-1}F)$ satisfies $(f, h_n) = (Dg, h_n) = -(g, h'_n) = 0$. But by the recurrence formula

$$(g, h'_n) = \sqrt{\frac{n}{2}}\,(g, h_{n-1}) - \sqrt{\frac{n+1}{2}}\,(g, h_{n+1}) = 0$$

or

$$(g, h_{n+1}) = \sqrt{\frac{n}{n+1}}\,(g, h_{n-1}), \qquad n = 0, 1, \cdots.$$

Thus g has the same Hermite coefficients as a constant function since $-(C, h_{n+1})\sqrt{\frac{n+1}{2}} + (C, h_{n-1})\sqrt{\frac{n}{2}} = 0$. By the induction hypothesis, since $g - C$ has zero coefficients, $g - C = 0$. Thus $Dg = 0$ and we are finished. □

THEOREM 6.7
Let $f \in S'$; then
    (i) $(f, h_n) = 0(n^p)$ *for some* $p \in \mathbb{Z}$,
    (ii) $\sum_n (f, h_n)h_n = f$ *in the sense of* S',
    (iii) *there is* $G \in L^2(\mathbb{R})$ *such that* $f = (x^2 - D^2)^r G$ *for some* $r \in \mathbb{Z}$.

PROOF   For (i), we use the fact $|(f, h_n)| \leq C \sup_t (1 + |t|)^r |h_n^{(r)}(t)|$ for some r (Chapter 2). But the recurrence formulae may each be applied r times to get first

$$\left| h_n^{(r)}(t) \right| \leq \sum_{k=-r}^{r} c_{n,k} |h_{n+k}(t)| \quad \text{where } c_{n,k} = 0(n^{r/2})$$

from (6.17b) and then

$$(1 + |t|)^r \left| h_n^{(r)}(t) \right| \leq \sum_{k=-2r}^{2r} d_{nk} |h_{n+k}(t)|, \quad d_{nk} = 0(n^r)$$

from (6.17a). Since $|h_n(t)|$ is uniformly bounded in t and n, the conclusion follows.

The series in (ii) clearly converges in S' since for $\phi \in S$,

$$(h_n, \phi) = (2n + 1)^{-q}((x^2 - D^2)^q h_n, \phi) = (2n + 1)^{-q}(h_n, (x^2 - D^2)^q \phi)$$

and $(x^2 - D^2)^q \phi \in L^2(\mathbb{R})$ for any $q = 0, 1, 2, \cdots$. By taking $q = p + 2$ we get the convergence of

$$\sum_n (f, h_n)(h_n, \phi).$$

If this series in (ii) converges to, say, g, then g and f both have the same Hermite coefficients and hence are the same.

The third conclusion follows from (i) by forming the series

$$\sum_n \frac{(f, h_n)}{(2n + 1)^{p+1}} h_n = G.$$

Since the coefficients are in ℓ^2, $G \in L^2(\mathbb{R})$. Then by applying the operator $(x^2 - D^2)^{p+1}$ to both sides we obtain

$$\sum_n (f, h_n) h_n = (x^2 - D^2)^{p+1} G.$$

Since this series also converges to f and $(x^2 - D^2)^{p+1} G \in S'$, they must be equal. $\square$

It should be noted that condition (i) and (iii) are also sufficient for $f \in S'$.

Another important property of Hermite functions is their relation to Fourier transforms. We first observe the well known fact that the Fourier transform of $e^{-x^2/2}$ is a multiple of the same function; in fact

$$\int_{-\infty}^{\infty} e^{-x^2/2} e^{-iwx} dx = \sqrt{2\pi}\, e^{-w^2/2}.$$

Thus, $\hat{h}_0(w) = \sqrt{2\pi}\, h_0(w)$, and for other h_n we use the fact that

$$\int_{-\infty}^{\infty} e^{-iwx}(x - D)\phi(x)dx = (-i)(w - D)\int_{-\infty}^{\infty} e^{-iwx}\phi(x)dx$$

for $\phi \in S$. Hence by taking $\phi = h_n$, we get by (6.18)

$$\int_{-\infty}^{\infty} e^{-iwx}\sqrt{2n+2}\,h_{n+1}(x)dx = (-i)(w - D)\hat{h}_n(w), \quad n = 0, 1, 2, \cdots,$$

or

$$\hat{h}_{n+1}(w) = (-i)\frac{1}{\sqrt{2n+2}}(w - D)\hat{h}_n(w).$$

Thus, for $n = 0$, we have $\hat{h}_1(w) = (-i)\frac{\sqrt{2\pi}}{\sqrt{2}}(w - D)h_0(w) = -i\sqrt{2\pi}\,h_1(w)$ and by induction

$$\hat{h}_n(w) = (-i)^n\sqrt{2\pi}\,h_n(w).$$

Thus the Hermite functions are eigenfunctions of the Fourier transform. This is the starting point for a simple theory of Fourier transforms.

PROPOSITION 6.8
Let $f \in L^2(\mathbb{R})$ (resp $\in S'$) with Hermite series

$$f \sim \sum_n a_n h_n;$$

then the Fourier transform $\hat{f}$ is given by

$$\hat{f}(w) \sim \sqrt{2\pi}\sum_n (-i)^n a_n h_n(w)$$

with convergence in the sense of $L^2(\mathbb{R})$ (resp S').

The proof is obvious. The reader may wish to explore the implications of this result.

6.3 Problems

1. Find a complete set of orthogonal polynomials on $(-\infty, \infty)$ with respect to $d\mu = \frac{1}{4}\delta(x - 1) + \frac{1}{4}\delta(x + 1) + \frac{1}{2}\delta(x)$.

2. Use the recurrence formula (6.8) for the Legendre polynomials to find $P_3(x)$ and $P_4(x)$, and show their orthogonality by direct calculation.

3. Show directly that the zeros of $P_3(x)$ and $P_4(x)$ are interlaced.

4. Find the expansion of the hat function $f(x) = 1 - |x|$ on $(-1, 1)$ with respect to Legendre polynomials.

5. The Legendre polynomials satisfy Rodrigues' formula

$$P_n(x) = \frac{1}{2^n n!} \frac{d^n (x^2 - 1)^n}{dx^n}.$$

   Use this to prove the orthogonality.

6. Use Rodrigues' formula to prove

$$P'_{n+1}(x) - P'_{n-1}(x) = (2n + 1) P_n(x).$$

7. Find the first five Tchebychev polynomials (Jacobi polynomials with $\alpha = \beta = \frac{1}{2}$) by using the recurrence formula (6.14).

8. Expand x^4 in terms of Tchebychev polynomials.

9. Find the first five Laguerre polynomials for $\alpha = 0$, and find the expansion of x^4 in terms of Laguerre polynomials.

10. Show that $L_n^{(0)}(x)$ satisfies the Rodrigues' formula

$$L_n^{(0)}(x) = (-1)^n \frac{d^n (x^n e^{-x})}{dx^n} e^x.$$

11. Prove the recurrence relation (6.15) for $\alpha = 0$ from the formula in Problem 10.

12. Expand $\delta(x)$ in terms of Hermite functions $\{h_n\}$. Use (6.17a) to establish a formula for the coefficients.

13. Expand $f(x) = 1$ in terms of Hermite functions. Use (6.17b) to establish a formula for the coefficients.

14. Show that $\hat{\delta}(\omega) = 1$ by combining Problems 12 and 13 and using Proposition 6.8.

15. Find the Hermite expansion of x and x^2 by using (6.17a) and (6.17b).

16. Prove Parseval's identity $\|f\|^2 = \frac{1}{2\pi} \|\hat{f}\|^2$ by using Proposition 6.8.

7

Other Orthogonal Systems

Orthogonal systems other than the trigonometric, orthogonal polynomials, and wavelets also play an important role in applications. Chief among these are the systems arising from <u>boundary value problems</u>, which are primarily used to solve partial differential equations. We present here only the barest essentials of these systems. A more complete (much more) description may be found in [T] or [C-L].

Another system consists of <u>eigenfunctions of integral operators</u>; these have proved to be useful in stochastic processes [R] as well as differential equations. These systems are related to Sturm-Liouville systems since the latter are also eigenfunctions of the integral operator with the associated Green's function [C-L]. In one case, that of the <u>prolate spheroidal functions</u>, the system is composed of eigenfunctions of both a differential operator and an integral operator which is not associated with the Green's function [Sl], [W3].

Two systems related to the Haar system, the Rademacher and Walsh systems, are also discussed in this chapter. The latter has been used for digital transmission of information (see [Ha], [O]).

Periodic wavelets that give orthonormal bases of finite intervals are included in this chapter. These are obtained by periodizing wavelets on the real line [D].

Finally, one of the predecessors to wavelets, the "localized sine basis" of Coifman and Meyer is constructed [C-M], [A-W-W]. This is an orthonormal version of a windowed Fourier transform.

7.1   Self adjoint eigenvalue problems on a finite interval

One of the classical techniques used to solve partial differential equations with associated boundary conditions is separation of variables, which converts the equation into ordinary differential equations. These equations also have certain boundary conditions and have solutions (eigenfunctions) only for certain values (eigenvalues) of a parameter.

A simple example is the wave equation

$$\frac{\partial^2 u}{\partial x^2} - \frac{\partial^2 u}{\partial t^2} = 0, \qquad u(0,t) = u(\pi,t) = 0,$$

$$0 \leq x \leq \pi, \ 0 \leq t.$$

The separation of variables consists of looking for solutions of the form

$$u(x,t) = f(x)g(t),$$

which must satisfy

$$f''g - fg'' = 0, \qquad f(0) = f(\pi) = 0. \tag{7.1}$$

Then (7.1) is written as

$$\frac{f''(x)}{f(x)} = \frac{g''(t)}{g(t)} = \lambda, \qquad f(0) = f(\pi) = 0,$$

where λ is a constant. The eigenvalue problem then is

$$y'' - \lambda y = 0, \qquad y(0) = y(\pi) = 0,$$

which $y = f$ must satisfy. The values of λ for which a solution exists are $\lambda_k = -k^2$, $k \in \mathbb{Z}$ and the resulting solutions are $y_k = \sin kx$. These y_k's are orthogonal on $(0, \pi)$ and are complete in $L^2(0, \pi)$. (This follows directly from considerations of Fourier series in Chapter 4.)

To solve the original equation one also has to find the $g(t)$, which in this case for $\lambda = -k^2$ is also of the form

$$g_k(t) = A_k \cos kt + B_k \sin kt.$$

Thus the general solution to the boundary value problem is

$$w(x,t) = \sum_{k=1}^{\infty} (A_k \cos kt + B_k \sin kt) \sin kx.$$

The coefficients A_k and B_k are found from initial conditions that do not concern us here.

The procedure can be imitated in a large number of other cases. We shall consider only the second order case on a finite interval (a, b). Let L be the differential operator

$$Ly = -(py')' + qy,$$

where $p \in C^1[a, b]$ and $q \in C[a, b]$. Then the problem is

$$Ly = \lambda y, \qquad \alpha_1 y(a) + \alpha_2 y'(a) = 0$$
$$\beta_1 y(b) + \beta_2 y'(b) = 0. \tag{7.2}$$

If p, q, α_1, α_2, β_1, and β_2 are real, and $p \neq 0$ on $[a, b]$, then (7.2) is a self adjoint, second order, linear eigenvalue problem commonly called a Sturm-Liouville problem. By self adjoint is meant the condition

$$\langle Lf, g \rangle = \langle f, Lg \rangle$$

for f and g satisfying the boundary conditions in (7.2). Such problems have solutions $\{\lambda_k, y_k\}$ such that the eigenfunctions $\{y_k\}$ are a complete orthonormal system in $L^2(a, b)$. See [C-L, p.197].

Indeed the same result is true for higher order linear differential operators and more general boundary values, but the Sturm-Liouville (S-L) problems are the most important in applications. We have described <u>regular</u> S-L problems that involve a finite interval on which $p \neq 0$. There are also <u>singular</u> S-L problems, which sometimes (but not always) have complete systems of eigenfunctions. These occur when the coefficients are singular, or $p(x) = 0$ for some $x \in [a, b]$, or the interval is infinite. Various standard orthogonal systems such as Hermite functions, Legendre polynomials, and Bessel functions arise as examples of solutions for such singular S-L problems. They are covered in great detail elsewhere and will not be discussed further here. See [T], [J].

A remarkable additional property of regular S-L problems is the existence of a <u>Green's function</u>. This is a function $g(t, s)$, $t, s \in [a, b]$, which forms the kernel of an integral operator that inverts the differential operator L (provided 0 is not an eigenvalue). It satisfies

$$L_t g(t, s) = \delta(t - s) \qquad t, s \in (a, b)$$

as well as the boundary conditions in (7.2) and is continuous on $[a, b] \times [a, b]$. The associated integral operator is

$$(Gf)(t) := \int_a^b g(t, s) f(s) ds$$

and is an inverse to L in the sense that if for $f \in [a, b]$, $Lw = f$ and w satisfies the boundary conditions, $w = Gf$. (See [C-L, p.192])

This operator G is a special case of a self adjoint compact operator, which we consider in the next section.

7.2   Hilbert-Schmidt integral operators

We here consider integral operators with kernels $q(t,s)$, $t,s \in (a,b)$, (a,b) finite, which belong to $L^2[(a,b) \times (a,b)]$ and which are symmetric, $q(t,s) = \overline{q(s,t)}$. The operator Q given by

$$(Qf)(t) := \int_a^b q(t,s)f(s)ds$$

is a <u>Hilbert-Schmidt</u> operator and clearly maps $L^2(a,b)$ into itself. These operators have been extensively studied and have the following properties [R-N, p.227] [R, p.203]:

(i)   There is at least one nonzero eigenvalue λ_1 with an eigenfunction $(Q\phi_1 = \lambda_1\phi_1)$.

(ii)   Each nonzero eigenvalue λ_n has finitely many independent eigenfunctions.

(iii)   The set of eigenfunctions $\{\phi_n\}$ corresponding to nonzero eigenvalues may be taken to be orthogonal.

(iv)   Each $g \in L^2(a,b)$ has a representation

$$g(t) = h(t) + \sum_n \langle g, \phi_n \rangle \phi_n(t)$$

where $Qh(t) = 0$.

(v) $g(t,s) = \sum_n \lambda_n \phi_n(t)\overline{\phi_n(s)}$.

Here the convergence is in the sense of $L^2(a,b)$ for s fixed. Since the symmetric Green's functions of the last section are particular cases, these operators are a rich source of orthogonal systems. However, their practical utility is limited by the difficulty of calculating the $\phi_n(t)$.

If Q is positive definite, i.e., if $\langle Qf, f \rangle > 0$ for $f \neq 0$, then (iv) reduces to

$$g(t) = \sum_n \langle g, \phi_n \rangle \phi_n(t),$$

i.e., $\{\phi_n\}$ is a complete orthonormal system. In this case the eigenvalues λ_n are all positive, and the $\phi_n(t)$ may be taken to be real valued functions (both the real and imaginary parts are eigenfunctions).

The eigenvalues and eigenfunctions may be approximated by means of Fourier series. By means of a translation and dilation if necessary, we may assume the interval is $(0, 2\pi)$. Then we have

$$q(t,s) = \sum_m \sum_k q_{mk} e^{imt} e^{-iks}$$

convergent in $L^2(0, 2\pi)^2$. If λ and $\phi(t)$ are an eigenvalue-eigenfunction pair, and $\phi(t)$ has Fourier series

$$\phi(t) = \sum_k c_k e^{ikt},$$

then

$$\int_0^{2\pi} q(t, s)\phi(s)ds = \sum_m \left(\sum_k q_{mk} c_k \right) e^{imt}$$

$$= \lambda \sum_m c_m e^{imt}.$$

The series $\sum_k q_{mk} c_k$ converges by Cauchy's inequality to a sequence in ℓ^2. Hence by truncating both series we have the algebraic eigenvalue problem

$$\sum_{|k| \leq N} q_{mk} c_k = \lambda c_m, \qquad |m| \leq N.$$

We then solve this for λ and $\{c_k\}$ to obtain an approximation to the true λ and $\phi(t)$. However, if the $q(t, s)$ is the Green's function of a differential operator, the properties of the latter can be used to obtain better approximations.

It would be nice if we could associate with Q a differential operator in cases when $q(t, s)$ is not a Green's function. This does occur in at least one case, but as we shall see in the next section, this is the only case. If each eigenvalue has multiplicity one, then we need only require that the differential operator L commute with Q. In this case we have

$$L Q\phi_n = L \lambda_n \phi_n$$

or

$$Q L\phi_n = \lambda_n L\phi_n$$

so that $L\phi_n$ is also an eigenfunction of Q with eigenvalue λ_n. Thus $L\phi_n$ must be some multiple of ϕ_n, i.e., ϕ_n is also an eigenfunction of L but possibly belonging to a different eigenvalue.

7.3 An anomaly — the prolate spheroidal functions

The prolate spheroidal (wave) functions $\{\phi_n\}$ are the eigenfunctions of the integral operator

$$(S_{\tau,\sigma}\phi)(t) := \int_{-\tau}^{\tau} \frac{\sin \sigma(t - x)}{\pi(t - x)} \phi(x)dx, \quad \tau, \sigma > 0. \tag{7.3}$$

The kernel is Hilbert-Schmidt on the interval $(-\tau, \tau)$, and thus the eigenfunctions may be taken to be orthonormal on $(-\tau, \tau)$. It is also positive definite so that $\{\phi_n\}$ is complete in $L^2(-\tau, \tau)$. By (7.3), they belong to the Paley-Wiener space of σ band limited functions since the kernel is exactly the reproducing kernel of B_σ (prototype II in Chapter 1).

We may write the eigenvalue problem in (7.3) as

$$\lambda\phi = (\chi_\tau \phi) \star \left(\frac{\sin \sigma t}{\pi t}\right)$$

and hence by taking Fourier transforms

$$\lambda\hat{\phi} = \widehat{(\chi_\tau \phi)} \cdot \chi_\sigma$$

since the Fourier transform of $(\sin \sigma t)/\pi t$ is the characteristic function χ_σ of $[-\sigma, \sigma]$. This last equation implies that the support of $\hat{\phi}$ is in $[-\sigma, \sigma]$, which therefore must satisfy

$$\hat{\phi} = \hat{\phi}\chi_\sigma.$$

By taking the inverse Fourier transform we see that

$$\phi = \phi \star \left(\sin \frac{\sigma t}{\pi t}\right)$$

or

$$\phi(t) = \int_{-\infty}^{\infty} \frac{\sin \sigma(t - x)}{\pi(t - x)}\phi(x)dx. \tag{7.4}$$

Thus we find that each ϕ_n is an eigenfunction of the same kernel but on the infinite interval $(-\infty, \infty)$ as well as $(-\tau, \tau)$. They all belong to the same eigenvalue 1 however. This is no contradiction since the kernel is not Hilbert-Schmidt on $(-\infty, \infty)$. A further surprising property is that the $\{\phi_n\}$ are orthogonal on $(-\infty, \infty)$ also. To see this we use (7.3) in

$$\int_{-\infty}^{\infty} \phi_n \phi_m = \int_{-\infty}^{\infty} \phi_n(t)\frac{1}{\lambda_m}\int_{-\tau}^{\tau} \phi_m(x)\frac{\sin \sigma(t - s)}{\pi(t - x)}dx\, dt$$

$$= \frac{1}{\lambda_m}\int_{-\tau}^{\tau} \phi_m(x)\int_{-\infty}^{\infty} \phi_n(t)\frac{\sin \sigma(t - x)}{\pi(t - x)}dt\, dx$$

$$= \frac{1}{\lambda_m}\int_{-\tau}^{\tau} \phi_m(x)\phi_n(x)dx = \frac{1}{\lambda_m}\delta_{mn}.$$

The interchange of the order of integration is justified since the integrand is absolutely integrable over the strip $[-\tau, \tau] \times (-\infty, \infty)$.

There are many more interesting properties of these $\{\phi_n\}$ that make them useful in signal analysis. (See [Pa, p.205] or [Sl].) But perhaps

the most interesting is the fact that they are also eigenfunctions of yet
a third operator.

7.4 A lucky accident?

The prolate spheroidal wave functions $\{\phi_n\}$ may also be characterized
as eigenfunctions of the differential operator [Sl]

$$P_{\tau,\sigma}\phi := (\tau^2 - t^2)\frac{d^2\phi}{dt^2} - 2t\frac{d\phi}{dt} - \sigma^2 t^2 \phi. \qquad (7.5)$$

Many of the useful properties of the $\{\phi_n\}$ derive from the fact that the
two operators commute

$$P_{\tau,\sigma}S_{\tau,\sigma}\phi = S_{\tau,\sigma}P_{\tau,\sigma}\phi$$

and hence that the $\{\phi_n\}$ are the eigenfunctions of either one. The first
shows they are entire functions of exponential type while the second is
used to determine local properties. This is the "lucky accident" referred
to by Slepian. However both operators are related to a third operator consisting of multiplication by the characteristic function χ_σ of an
interval $[-\sigma, \sigma]$ and to the Fourier transform

$$(\mathcal{F}\phi)(w) = \frac{1}{\sqrt{2\pi}} \int_{-\infty}^{\infty} \phi(t)e^{-iwt}dt = \hat{\phi}(w).$$

Here we have added the factor $\frac{1}{\sqrt{2\pi}}$ in order to obtain symmetric formulae for both the Fourier transform and its inverse. We do this only
in this section. We shall show that (7.5) is the unique differential operator such that both it and its Fourier transform are second order and
commute with characteristic functions.

   Let $P(x, D)$ be the linear second order differential operator with polynomial coefficients given by

$$P(x, D)\phi := \rho_0(x)D^2\phi + \rho_1(x)D\phi + \rho_2(x)\phi.$$

It is easy to derive conditions under which $P(x, D)$ commutes with multiplication by the characteristic functions χ_σ (with differentiation in the
sense of distributions) as given in the following lemma.

LEMMA 7.1
$P(x, D)(\chi_\sigma(x)\phi(x)) = \chi_\sigma(x)P(x, D)\phi(x)$ *for all* $\phi \in C^2$ *if and only if*

$$\rho_0(\pm\sigma) = 0, \qquad \rho_1(\pm\sigma) = \rho_0'(\pm\sigma).$$

PROOF By direct computation we have

$$P(\cdot, D)(\chi_\sigma \phi) = \rho_0(\chi''_\sigma \phi + 2\chi'_\sigma \phi') + \rho_1 \chi'_\sigma \phi + \dot{\chi}_\sigma P(\cdot, D)\phi$$
$$= \rho_0 \phi(\delta'_{-\sigma} - \delta'_\sigma) + (2\rho_0 \phi' + \rho_1 \phi)(\delta_{-\sigma} - \delta_\sigma)$$
$$+ \chi_\sigma P(\cdot, D)\phi.$$

Hence the two operators commute if and only if

$$(\rho_0 \phi)(\delta'_{-\sigma} - \delta'_\sigma) + (2\rho_0 \phi' + \rho_1 \phi)(\delta_{-\sigma} - \delta_\sigma) = 0 \qquad (7.6)$$

for all ϕ. In particular if ϕ has support on $(0, \infty)$ and (7.6) holds we see that

$$\rho_0 \phi \delta'_\sigma + (2\rho_0 \phi' + \rho_1 \phi)\delta_\sigma = 0$$

and since $f\delta'_\sigma = f(\sigma)\delta'_\sigma - f'(\sigma)\delta_\sigma$, $f\delta_\sigma = f(\sigma)\delta_\sigma$, it follows that

$$\rho_0(\sigma)\phi(\sigma)\delta'_\sigma - (\rho'_0(\sigma)\phi(\sigma) + \rho_0(\sigma)\phi'(\sigma) - 2\rho_0(\sigma)\phi'(\sigma)$$
$$- \rho_1(\sigma)\phi(\sigma))\delta_\sigma = 0.$$

Hence if $\phi(\sigma) \neq 0$, $\rho_0(\sigma) = 0$ and $\rho'_0(\sigma) - \rho_1(\sigma) = 0$. A similar argument works for $-\sigma$.

These conditions are also clearly sufficient for (7.6).   □

Example 1
The operator $P_{\tau,\sigma}$ of (7.5) commutes with χ_τ.

Example 2
The operator of the Jacobi polynomials $\{P_n^{(\alpha,\beta)}\}$ [Chapter 6]

$$Py = (1 - x^2)D^2 y + [(\beta - \alpha) - (\alpha + \beta + 2)x]Dy$$

commutes with χ_1 if and only if $\alpha = \beta = 0$, i.e. the Legendre case.

Example 3
The operator of the Laguerre polynomials $\{L_n^\alpha\}$

$$Py = xD^2 y + (\alpha + 1 - x)Dy$$

commutes with the characteristic function of $[0, \infty)$ if and only if $\alpha = 0$.

If the operator $P(x, D)$ has quadratic polynomial coefficients, then it assumes a particular simple form under the hypothesis of Lemma 7.1.

COROLLARY 7.2

Let $\rho_0(x)$ and $\rho_1(x)$ be quadratic polynomials; then $P(x, D)$ commutes with χ_σ if and only if

$$\rho_0(x) = a(x^2 - \sigma^2), \qquad \rho_1(x) = 2ax + b(x^2 - \sigma^2)$$

for some constant a and b.

The Fourier transform maps differential operators $P(x, D)$ of the form (7.5) into operators $P(iD, ix)$. That is

$$(P(\widehat{\cdot, D})\phi)(w) = P(iD, iw)\hat{\phi}(w).$$

In particular if $P(x, D)$ has quadratic coefficients, then $P(iD, ix)$ is also a second order linear operator with quadratic coefficients. By Corollary 7.2 those $P(x, D)$ which commute with χ_σ have the form

$$P(x, D) = [x^2 \ \ x \ \ 1] \begin{bmatrix} a & b & c_1 \\ 0 & 2a & c_2 \\ -a\sigma^2 & -b\sigma^2 & c_2 \end{bmatrix} \begin{bmatrix} D^2 \\ D \\ 1 \end{bmatrix}.$$

Clearly $P(iD, ix)$ has a similar form

$$P(iD, ix) = [x^2 \ \ x \ \ 1] \begin{bmatrix} a & 0 & -a\sigma^2 \\ -ib & 2a & ib\sigma^2 \\ -c_1 & i(c_2 - 2b) & c_3 \end{bmatrix} \begin{bmatrix} D^2 \\ D \\ 1 \end{bmatrix}.$$

This is shown by the calculation

$$\begin{aligned} P(iD, ix) &= a(D^2 - \sigma^2)x^2 - ib(D^2 - \sigma^2)x - 2aDx - c_1D^2 + ic_2D + c_3 \\ &= a(x^2D^2 + 4xD + 2 - \sigma^2x^2) - ib(xD^2 + 2D - \sigma^2x) \\ &\quad - 2a(xD + 1) - c_1D^2 + ic_2D + c_3. \end{aligned}$$

If in addition $P(iD, ix)$ commutes with χ_τ, then the expression is further simplified since Lemma 7.1 may be applied to both operators. This gives us

LEMMA 7.3

If $P(x, D)$ commutes with χ_σ and $P(iD, ix)$ commutes with χ_τ, then

$$P(x, D) = a(x^2 - \sigma^2)D^2 + 2axD + a\tau^2x^2 + c$$
$$P(iD, ix) = a(x^2 - \tau^2)D^2 + 2axD + a\sigma^2x^2 + c$$

and conversely.

The operator $P_{\tau,\sigma}$ (7.5) is an operator of the form considered here. Its Fourier transform operator is $P_{\sigma,\tau}(x,D) = P_{\tau,\sigma}(iD, ix)$. Hence $P_{\tau,\sigma}$ commutes with χ_τ and $P_{\sigma,\tau}$ with χ_σ.

The integral operator $S_{\tau,\sigma}$ of (7.3) may be expressed as

$$S_{\tau,\sigma}\phi = \sqrt{2\pi}\,\hat\chi_\sigma * (\chi_\tau\phi).$$

Hence we have

COROLLARY 7.4

$$P_{\tau,\sigma}(\hat\chi_\sigma * (\chi_\tau\phi)) = \hat\chi_\sigma * (\chi_\tau P_{\tau,\sigma}\phi). \tag{7.7}$$

PROOF The conjugate (inverse) Fourier transform of the left side of (7.7) is
$$P_{\sigma,\tau}(\chi_\sigma(\tilde\chi_\tau * \tilde\phi)) = \chi_\sigma P_{\sigma,\tau}(\tilde\chi_\tau * \tilde\phi).$$

By taking the Fourier transform of the right side of this expression, we obtain
$$\hat\chi_\sigma * P_{\tau,\sigma}(\chi_\tau\phi) = \hat\chi_\sigma * (\chi_\tau P_{\tau,\sigma}\phi)$$

which is the right side of (7.7).   □

COROLLARY 7.5
Let $P(x,D)$ be a linear second order differential operator with quadratic coefficients such that $P(x,D)$ commutes with χ_σ, and $P(iD,ix)$ commutes with χ_τ; then
$$P(x,D) = aP_{\sigma,\tau} + b$$

for constants a and b.

This is merely a restatement of Lemma 7.3.

We can come full circle by showing that if Corollary 7.4 holds for an operator $P(x,D)$ then the hypothesis of Corollary 7.5 is satisfied.

THEOREM 7.6
Let $P = P(x,D)$ be a second order linear differential operator with quadratic coefficients; then P commutes with the integral operator $S_{\tau,\sigma}$, if and only if
$$P = aP_{\tau,\sigma} + b.$$

PROOF We must show that $P(x,D)$ commutes with χ_σ and $P(iD,ix)$ with χ_τ. The commutator of $S_{\tau,\sigma}$ and P applied to ϕ is

$$P(\hat\chi_\sigma * (\chi_\tau\phi)) - \hat\chi_\sigma * (\chi_\tau P\phi) = 0.$$

That of P and χ_τ is

$$P(\chi_\tau \phi) - \chi_\tau P\phi = a_{-\tau}\delta'_{-\tau} + a_\tau \delta'_\tau + b_{-\tau}\delta_{-\tau} + b_\tau \delta_\tau \qquad (7.8)$$

for some constants $a_{-\tau}$, a_τ, $b_{-\tau}$ and b_τ by calculations similar to those in Lemma 7.1. Similarly the commutator of $\hat{P} = P(iD, ix)$ and χ_σ has the same form

$$\hat{P}(\chi_\sigma \psi) - \chi_\sigma \hat{P}\psi = c_{-\sigma}\delta'_{-\sigma} + c_\sigma \delta'_\sigma + d_{-\sigma}\delta_{-\sigma} + d_\sigma \delta_{-\sigma}. \qquad (7.9)$$

We now express the difference of the two sides of (7.7) as

$$\begin{aligned} PS_{\tau,\sigma}\phi - S_{\tau,\sigma}P\phi &= P(\hat{\chi}_\sigma * (\chi_\tau \phi)) - \hat{\chi}_\sigma * P(\chi_\tau \phi) \\ &\quad + \hat{\chi}_\sigma * P(\chi_\tau \phi) - \hat{\chi}_\sigma * (\chi_\tau P\phi) \\ &= \mathcal{F}^{-1}\left[(\hat{P}\chi_\sigma - \chi_\sigma \hat{P})(\hat{\chi}_\tau * \hat{\phi})\right] \\ &\quad + \hat{\chi}_\sigma * (P(\chi_\tau \phi) - \chi_\tau P\phi) \end{aligned} \qquad (7.10)$$

where $\mathcal{F}^{-1}$ denotes the inverse Fourier transform and we have used the fact that the Fourier transform of $\hat{\chi}_\sigma$ is χ_σ. Therefore (7.10) is a linear combination of the eight functions obtained by substituting (7.8) and (7.9) in (7.10).

These in turn are linear combinations of

$$te^{-i\sigma t}, \; te^{i\sigma t}, \; e^{-i\sigma t}, \; e^{i\sigma t}, \; \frac{\sin \sigma(t-\tau)}{(t-\tau)}, \; \frac{\sin \sigma(t+\tau)}{(t+\tau)},$$

$$\frac{(t-\tau)\cos \sigma(t-\tau) - \sin \sigma(t-\tau)}{(t-\tau)^2},$$

$$\frac{(t+\tau)\cos \sigma(t+\tau) - \sin \sigma(t+\tau)}{(t+\tau)^2},$$

which are linearly independent whenever $\sigma\tau \neq 0$. Since (7.10) is zero for all ϕ, it follows that all the coefficients must be zero. Hence (7.8) and (7.9) must be zero, and the hypothesis of Corollary 7.5 is satisfied. $\square$

Thus this is indeed a "lucky accident".

7.5 Rademacher functions

The Haar functions, which constituted the first prototype example of wavelets, are orthogonal on $(-\infty, \infty)$ if one takes both m and n ranging over $\mathbb{Z}$ in $\psi_{mn}(t) = 2^{m/2}\psi(2^m t - n)$. However a subset together with $\phi(t)$ constitute an orthogonal system in $[0,1]$ complete in $L^2(0,1)$. This subset consists of all $\psi_{mn}(t)$ with support contained in $[0, 1]$, i.e., all $m \geq 0$ and $0 \leq n \leq 2^m - 1$. The <u>Rademacher functions</u> are an orthogonal system on $[0, 1]$ obtained by adding up all the Haar functions at the same scale, i.e.,

$$r_0(t) := \phi(t),$$
$$r_1(t) := \psi(t),$$
$$r_{m+1}(t) := \sum_{n=0}^{2^m-1} \psi(2^m t - n).$$

These are clearly orthogonal because of the orthogonality of the subspaces $V_0, W_0, W_1, \cdots, W_m, \cdots$ to which $r_0, r_1, \cdots, r_{m+1}, \cdots$ belong. These are also not complete since $\langle \psi_1, r_k \rangle = 0$ for all $k \neq 2$ but $\psi_1 \neq r_0$.

Their principle use is to study series of the form

$$\sum_{n=0}^{\infty} \pm c_n \phi_n(x) \tag{7.11}$$

where the $+$'s and $-$'s are chosen at random. This is possible to rewrite as

$$\sum_{n=0}^{\infty} c_n r_n(t) \phi_n(x) \tag{7.12}$$

since, except for dyadic rational numbers, $r_n(t) = \pm 1$. In fact, if t has an infinite dyadic expansion

$$t = 0. \, a_1 a_2 a_3 \cdots$$

and the $+$'s correspond $a_n = 1$ and the $-$'s to $a_n = 0$, then (7.11) and (7.12) are the same. (The sign of c_0 is always taken to be $+$).

The principal result [Al, p.53] is that (7.11) is convergent almost everywhere for almost all choices of sign when $c_n \in \ell^2$. There are a number of other results related to probability, which may be found in [Z, p.212].

7.6   Walsh functions

The Rademacher functions were obtained by combining the Haar functions by simply adding them at a given scale. The Walsh functions take sums and differences of the Haar functions to obtain a complete system.

We define

$$w_0(t) := \phi(t), \ w_1(t) := \psi(t),$$
$$w_2(t) := \psi(2t) + \psi(2t - 1),$$
$$w_3(t) := \psi(2t) - \psi(2t - 1),$$

$$\cdots \tag{7.13}$$

$$w_{2n}(t) := w_n(2t) + w_n(2t - 1),$$
$$w_{2n+1}(t) := w_n(2t) - w_n(2t - 1),$$

$$\cdots .$$

Thus these Walsh functions also belong to the wavelet subspaces of the Haar system:

$$w_0 \in V_0, \ w_1 \in W_0, \ w_2, w_3 \in W_1, \ w_4, w_5, w_6, w_7 \in W_2, \cdots$$

$$w_{2^m}, \ w_{2^m+1}, \cdots, w_{2^{m+1}-1} \in W_m, \cdots .$$

Notice that these defining relations (7.13) are exactly the same as those in the two dilation equations of the Haar system,

$$\phi(t) = \phi(2t) + \phi(2t - 1), \tag{7.14a}$$

$$\psi(t) = \phi(2t) - \phi(2t - 1). \tag{7.14b}$$

Since all functions defined by (7.14a) are orthogonal to all defined by (7.14b), it follows that w_{2n} and w_{2n+1} are orthogonal. Also if w_n and w_m are orthogonal so are w_{2n} and w_{2m}. Hence by an induction argument $w_{2^m}, w_{2^m+1}, \cdots, w_{2^{m+1}-1}$ are orthogonal in W_m. Since all of these functions have support contained in $[0,1]$, the $[w_n]$ are an orthogonal system in $L^2(0,1)$. Moreover, there are exactly 2^m Haar functions in W_m whose support lies in $[0,1]$, and therefore the Walsh functions in W_m form a basis of this space. Since the Haar functions are complete in $L^2(0,1)$ so are the Walsh functions.

These Walsh functions play an important role in communications theory in network analysis and synthesis. (See, e.g., [Ha]). They also serve as a model for some of the wavelet packets of Coifman, Meyer, and Wickerhauser [C-M-W].

7.7 Periodic wavelets

Up to now all the wavelets we have considered are bases of $L^2(\mathbb{R})$. However, in many applied problems, one is more interested in orthonormal systems on a finite interval. One such system can be found by periodizing the scaling functions and wavelets [D, p.304].

Let $\phi \in S_r$ be a scaling function with associated MRA $\{V_m\}$ and mother wavelet ψ. We define ϕ_{mn}^* and ψ_{mn}^* as the periodized versions of ϕ_{mn} and ψ_{mn},

$$\phi_{mn}^*(t) := \sum_k \phi_{mn}(t-k),$$

$$\psi_{mn}^*(t) := \sum_k \psi_{mn}(t-k).$$

Both are clearly periodic of period 1, and

$$\phi_{00}^*(t) = \sum_k \phi(t-k) = 1.$$

The same is true for $\phi_{mn}^*(t)$ for $m < 0$ since e.g.,

$$\phi_{-1,n}^*(t) = \sum_k 2^{-1/2} \phi\left(\frac{t}{2} - n - \frac{k}{2}\right)$$

$$= 2^{-1/2}\left(\sum_j \phi\left(\frac{t}{2} - n - j\right) + \sum_j \phi\left(\frac{t}{2} - n - \frac{1}{2} - j\right)\right)$$

$$= 2^{-1/2}(1+1) = 2^{1/2},$$

and $\phi_{m,n}^*(t)$ is in the space spanned by $\phi_{00}^*(t)$. Thus, the $\phi_{mn}^*(t)$ need only be considered for $m \geq 0$. We denote by V_m^* the space spanned by $\{\phi_{m,n}^*\}_{n \in \mathbb{Z}}$. This is now a finite dimensional space since $\phi_{m,n}^* = \phi_{m,n+2^m}^*$. In fact, $\phi_{m,0}^*, \phi_{m,1}^*, \cdots, \phi_{m,2^m-1}^*$ are an orthogonal basis in the sense of $L^2(0,1)$ of V_m^* since

$$\int_0^1 \phi_{m0}^*, \phi_{mn}^* = \int_0^1 \sum_k \phi(2^m(t-k)) \sum_j \phi(2^m(t-j)-n) 2^m \, dt$$

$$= \sum_j 2^m \sum_k \int_k^{k+1} \phi(2^m t)\phi(2^m(t-j+k)-n) dt$$

$$= 2^m \sum_k \int_k^{k+1} \phi(2^m t) \sum_j \phi(2^m(t-j)-n) dt$$

$$= \sum_j 2^m \int_{-\infty}^{\infty} \phi(2^m t)\phi(2^m(t-j)-n) dt = 0, \ 0 < n < 2^m.$$

These V_m^*'s have the nested property

$$V_0^* \subset V_1^* \subset \cdots \subset V_m^* \subset \cdots \subset L^2(0,1)$$

and
$$\overline{\bigcup_m V_m^*} = L^2(0, 1).$$

Again the V_m^*'s are RKHS with RK q_m^* given by

$$q_m^*(x, t) := \sum_{n=0}^{2^m - 1} \phi_{mn}^*(x)\phi_{mn}^*(t), \quad x, t \in [0, 1].$$

By using the result for $q_m(x, t)$, it can be shown that $\{q_m^*(x, t)\chi_{[0,1]}(x)\,\chi_{[0,1]}(t)\}$ is a quasi-positive delta sequence. Hence we have

PROPOSITION 7.7
Let $f(t)$ be continuous and periodic, then the projection f_m^* of f onto V_m^* converges to f uniformly.

The corresponding wavelets themselves are considerably simpler in the periodic case as well. If we denote by ψ_0 the constant function

$$\psi_0(t) := \phi_{00}^*(t) = 1$$

and by ψ_n the functions

$$\psi_{2^m + k}(t) = \psi_{m,k}^*(t), \quad 0 \le m, \ k = 0, 1, \cdots 2^m - 1,$$

then the $\{\psi_n\}$ are an orthonormal basis of $L^2(0, 1)$ consisting of periodic functions. It is clear that

$$\{\psi_0, \psi_1, \cdots, \psi_{2^m - 1}\}$$

is also an orthonormal basis of V_m^* since there are 2^m of them, and they are orthonormal. The orthogonal expansion of an $f \in L^2(0, 1)$ given by

$$f \sim \sum_{n=0}^{\infty} \langle f, \psi_n \rangle \psi_n$$

has as its 2^mth partial sum f_m^*. The other partial sums can be squeezed between f_m^* and f_{m+1}^* for some m, and hence for f continuous and periodic, the series converges to f uniformly. As we observed in Chapter 4, this is not true for trigonometric Fourier series.

7.8 Local sine or cosine bases

A number of procedures have been proposed for using the windowed Fourier transform to construct an orthonormal basis of $L^2(\mathbb{R})$. This is given by

$$\int_{-\infty}^{\infty} f(t)e^{iakt}w(t-bn)dt, \qquad n,k \in \mathbb{Z}$$

where $w(t)$ is the window. In general these functions $\{e^{iakt}w(t-bn)\}$ will not be orthogonal. However if $a=1$ and $b=2\pi$ with $w(t)$ given by the characteristic function of $[0,2\pi)$, $\chi(t)$, it is easy to see that

$$u_{k,n}(t) \ := \ \frac{e^{ikt}}{\sqrt{2\pi}}\chi(t-2\pi n)$$

is orthonormal in $L^2(\mathbb{R})$. Indeed, the restriction $f_n(t)$ of $f(t)$ to the interval $[2\pi n, 2\pi(n+1)]$ has a Fourier series convergent in the L^2 sense to the periodic extenison of $f_n(t)$. However, even if f is C^∞ this Fourier series will converge very slowly unless $f(2\pi(n+1))$ and $f(2\pi n)$ match up, and we will also get Gibbs' phenomenon at these end points.

This is only one of the shortcomings of this basis. The other is that the Fourier transform of $u_{k,n}(t)$ given by

$$\hat{u}_{k,n}(w) = \frac{e^{-i(w-k)n}}{\sqrt{2\pi}}\left\{\frac{e^{-i(w-k)2\pi}-1}{i(k-w)}\right\}$$

decreases very slowly as $|k| \to \infty$. To overcome this we must use a smooth window $w(t)$. But then it is difficult to obtain orthonormality. A number of procedures have been proposed. We present one, the local sine and cosine bases of Coifman and Meyer [C-M]. A detailed exposition of the construction may be found in [A-W-W].

The construction begins with a bell $b(x)$ similar to that used in the construction of the Meyer Wavelets in Chapter 3. Let $h(x)$ be a probability function (or distribution) with support on $[-1,1)$ and $h_\epsilon, \epsilon > 0$, be its dilation $h_\epsilon(x) = \frac{1}{\epsilon}h(x/\epsilon)$. Then the (ϵ, ϵ') bell over $[\alpha, \beta]$ is defined to be the positive square root of

$$b_{\alpha,\beta}^2(x) \ := \ \int_{-\epsilon}^{x-\alpha} h_\epsilon - \int_{-\epsilon'}^{x-\beta} h_{\epsilon'}, \tag{7.15}$$

which depends on α, β, ϵ, and ϵ'. If $\epsilon = \epsilon'$ this reduces to the expression

$$b^2(x) = \int_{x-\beta}^{x-\alpha} h_\epsilon.$$

It clearly has support on $[\alpha - \epsilon, \beta + \epsilon']$ and $b(x)b(2\alpha - x)$ has support on $[\alpha - \epsilon, \beta + \epsilon]$. Moreover, $b^2(x) + b^2(2\alpha - x) = 1$ for $2\alpha - \beta + \epsilon' < x < \beta - \epsilon'$, and similar results hold near β as well.

Two adjacent intervals $[\alpha, \beta]$ and $[\beta, \gamma]$ have compatible bells if the (ϵ, ϵ') bell over $[\alpha, \beta]$ and the (ϵ', ϵ'') bell over $[\beta, \gamma]$ satisfy $\gamma - \epsilon'' \geq \beta + \epsilon'$.

We can associate with the bell over $[\alpha, \beta]$ a pair of projection operators $P^\pm = P^\pm_{[\alpha, \beta]}$ on $L^2(\mathbb{R})$ given by

$$(P^\pm f)(x) := b(x)\{b(x)f(x) \pm b(2\alpha - x)f(2\alpha - x) \\ \mp b(2\beta - x)f(2\beta - x)\}. \qquad (7.16)$$

The properties of $b(x)$ transform into properties of these operators. Indeed we have

(i) $\quad P^{\pm 2} = P^\pm \quad$ (so that $P^\pm$ is a projection)

(ii) $\quad P^\pm_{[\alpha, \beta]} + P^\pm_{[\beta, \gamma]} = P^\pm_{[\alpha, \gamma]}$

(iii) $\quad P^\pm_{[\alpha, \beta]} P^\pm_{[\beta, \gamma]} = P^\pm_{[\beta, \gamma]} P^\pm_{[\alpha, \beta]} = 0 \quad$ (orthogonality).

We now take an increasing sequence $\{a_k\}$ and a sequence $\{\epsilon_k\}$ of positive numbers such that $a_{\pm k} \to \pm\infty$ and

$$a_k + \epsilon_k < a_{k+1} - \epsilon_{k+1}, \qquad k \in \mathbb{Z}. \qquad (7.17)$$

Then for each interval $[a_k, a_{k+1}]$ we find the $(\epsilon_k, \epsilon_{k-1})$ bell $b_k(x)$ and form the projectors $P^\pm_k$. Then these projectors are mutually orthogonal and

$$L^2(\mathbb{R}) = \bigoplus_k P^+_k(L^2(\mathbb{R})) \quad (P^-_k \text{ respectively}).$$

See [A-W-W] for more details.

Thus if we can find an orthogonal basis of the Hilbert spaces $\mathcal{H}^\pm = P^\pm L^2(\mathbb{R})$ we can get an orthogonal basis of all of $L^2(\mathbb{R})$. For simplicity we take the interval $[\alpha, \beta]$ to be $[0, 1]$. Then

$$\left\{ \sqrt{2}\, b(x) \cos\left(k + \frac{1}{2}\right)\pi x \right\}_{k=0}^{\infty} \left(\text{resp.} \left\{ \sqrt{2}\, b(x) \sin\left(k + \frac{1}{2}\right)\pi x \right\}_{k=0}^{\infty} \right)$$

is an orthonormal basis of $\mathcal{H}^+$ (resp. $\mathcal{H}^-$).

It is easy to show that $\{\sqrt{2} \cos\left(k + \frac{1}{2}\right)\pi x\}$ is an orthonormal basis of $L^2(0, 1)$ (Chapter 4). Hence on $[0, 1]$,

$$g(x) := b(x)f(x) + b(-x)f(-x) - b(2 - x)f(2 - x) \\ = \sum_{k=0}^{\infty} \sqrt{2}\, c_k \cos\left(k + \frac{1}{2}\right)\pi x$$

where $f \in L^2(\mathbb{R})$ with convergence in $L^2[0,1]$. Hence on $[0,1]$

$$(P^+f)(x) = b(x)g(x) = \sum_{k=0}^{\infty} \sqrt{2}\, c_k b(x) \cos\left(k + \frac{1}{2}\right)\pi x.$$

This can be extended to $[-\epsilon, 1 + \epsilon']$ since both $g(x)$ and $\cos\left(k + \frac{1}{2}\right)\pi x$ are even about $x = 0$ and odd about $x = 1$.

It remains to be shown that $\{w_k\}$, where $w_k(x) = \sqrt{2}\, b(x) \cos\left(k + \frac{1}{2}\right)\pi x$, is orthogonal in $\mathcal{H}^+ = P^+ L^2(\mathbb{R})$. By direct calculation

$$\int_{-\epsilon}^{1+\epsilon'} w_k(x)w_j(x)dx = \int_{-\epsilon}^{\epsilon} + \int_{\epsilon}^{1-\epsilon'} + \int_{1-\epsilon'}^{1+\epsilon'}$$

$$= 2\int_0^{\epsilon} \{b^2(x) + b^2(-x)\} \cos\left(k + \frac{1}{2}\right)\pi x \cos\left(j + \frac{1}{2}\right)\pi x\, dx$$

$$+ 2\int_{\epsilon}^{1-\epsilon'} \cos\left(k + \frac{1}{2}\right)\pi x \cos\left(j + \frac{1}{2}\right)\pi x\, dx$$

$$+ 2\int_{1-\epsilon'}^{1} (b^2(x) = b^2(2-x)) \cos\left(k + \frac{1}{2}\right)\pi x \cos\left(j + \frac{1}{2}\right)\pi x\, dx$$

$$= 2\int_0^{1} \cos\left(k + \frac{1}{2}\right)\pi x \cos\left(j + \frac{1}{2}\right)\pi x\, dx = \delta_{kj}$$

since $b^2(x) + b^2(-x) = 1$ as $[0, \epsilon]$ and $b^2(x) = b^2(2-x) = 1$ on $[1-\epsilon', 1]$.

Now by shifting the interval from $[0,1]$ to $[a_k, a_{k+1}]$ we can obtain an orthonormal basis of $\mathcal{H}_k^+ = P_k^+(L^2(\mathbb{R}))$ and hence of $L^2(\mathbb{R})$, and similarly for $\mathcal{H}_k^-$.

THEOREM 7.8
Let $\{a_k\}$ be a sequence of real numbers and $\{\epsilon_k\}$ of positive numbers such that $a_{\pm k} \to \pm\infty$ and

$$a_k + \epsilon_k < a_{k+1} - \epsilon_{k+1};$$

let $b_k(x)$ be the $(\epsilon_k, \epsilon_{k+1})$ bell over $[a_k, a_{k+1}]$; then $\{u_{k,j}\}$ where

$$u_{k,j}(x) = \sqrt{\frac{2}{a_{k+1} - a_k}}\, b_k(x) \cos\frac{(2j+1)\pi(x - a_k)}{2(a_{k+1} - a_k)},$$

$$k = 0, \pm 1, \pm 2, \cdots, \; j = 0, 1, 2, \cdots$$

and $\{v_{k,j}\}$ where

$$v_{k,j}(x) = \sqrt{\frac{2}{a_{k+1} - a_k}}\, b_k(x) \sin\frac{(2j+1)\pi(x - a_k)}{2(a_{k+1} - a_k)},$$

$$k = 0, \pm 1, \pm 2, \cdots, \; j = 0, 1, 2, \cdots$$

are each an orthonormal basis of $L^2(\mathbb{R})$.

These are very versatile systems since the $\{a_k\}$ are quite general. If it is known that a signal changes its frequency structure at points $\{a_k\}$, then these are points to use. There is also a "merging" algorithm [A-W-W] that enables one to choose an optimal sequence $\{a_k\}$ by removing points successively.

7.9 Biorthogonal wavelets

Each of the techniques used in Chapter 3 to construct orthogonal scaling functions and consequently orthogonal wavelets may be modified to construct biorthogonal pairs of scaling functions. Such constructions allow greater flexibility but at the cost of greater complexity.

The biorthogonal pair will be denoted as $(\phi, \tilde{\phi})$ where each is a function (or distribution) on $\mathbb{R}$ and satisfies

$$\int \phi(t)\overline{\tilde{\phi}(t - n)}dt = \delta_{on} \tag{7.18}$$

We expect that $\{\phi(t - n)\}$ (resp. $\{\tilde{\phi}(t - n)\}$) is a Riesz basis of its closed linear span V_0 (resp. $\tilde{V}_0$). This leads to a pair of multiresolution analyses,

$$\cdots \subset V_{-1} \subset V_0 \subset V_1 \subset \cdots$$
$$\cdots \subset \tilde{V}_{-1} \subset \tilde{V}_0 \subset \tilde{V}_1 \subset \cdots,$$

a pair of dilation equations,

$$\phi(t) = \sqrt{2} \sum_n h_n \phi(2t - n)$$
$$\tilde{\phi}(t) = \sqrt{2} \sum_n \tilde{h}_n \tilde{\phi}(2t - n)$$
$$\left[\hat{\phi}(w) = m_0(\frac{w}{2})\hat{\phi}(\frac{w}{2}) \right.$$
$$\left. \hat{\tilde{\phi}}(w) = \tilde{m}_0(\frac{w}{2})\hat{\tilde{\phi}}(\frac{w}{2}) \right], \tag{7.19}$$

and a pair of mother wavelets

$$\psi(t) = \sqrt{2} \sum_n \tilde{h}_{-n-1}(-1)^n \phi(2t - n)$$
$$\tilde{\psi}(t) = \sqrt{2} \sum_n h_{-n-1}(-1)^n \tilde{\phi}(2t - n). \tag{7.20}$$

We would like to place conditions on ϕ and $\tilde{\phi}$ such that each $f \in L^2(\mathbb{R})$ has an expansion

$$f = \sum \langle f, \tilde{\psi}_{kj} \rangle \psi_{kj} = \sum \langle f, \psi_{kj} \rangle \tilde{\psi}_{kj},$$

i.e., that $\{\psi_{kj}\}$ and $\{\tilde{\psi}_{kj}\}$ be dual Riesz bases of $L^2(\mathbb{R})$.

There are many possible procedures for choosing ϕ and $\tilde{\phi}$ in order to obtain such a result. The simplest is the one used to construct the Franklin wavelets in Chapter 3. It begins with the hat function

$$\phi(t) := (1 - |t|)\chi_{[-1,1]}(t),$$

which is not orthogonal to its translates. However $\{\phi(t-n)\}$ is a Riesz basis of V_0. Its Fourier transform is found to be

$$\hat{\phi}(w) = \left[\frac{\sin w/2}{w/2}\right]^2.$$

We must choose $\tilde{\phi}$ such that (7.18) is satisfied or in terms of the Fourier transform such that

$$\sum_k \hat{\phi}(w + 2\pi k)\,\overline{\hat{\tilde{\phi}}(w + 2\pi k)} = 1. \qquad (7.21)$$

We substitute $\hat{\phi}(w)$ into (7.21) and solve for $\tilde{\phi} \in V_0$, i.e.,

$$\hat{\tilde{\phi}}(w) = \sum_k a_k e^{-iwk}\hat{\phi}(w) = a(w)\hat{\phi}(w).$$

Then (7.21) becomes

$$\sum_k |\hat{\phi}(w + 2\pi, k)|^2 \overline{a(w)} = 1,$$

and since we know the sum of this series (section 3.1, example 2), it follows that

$$a(w) = \frac{1}{1 - \frac{2}{3}\sin^2 \frac{w}{2}}.$$

Hence $\tilde{\phi}$ is given by

$$\hat{\tilde{\phi}}(w) = \frac{\sin^2 w/2}{(1 - \frac{2}{3}\sin^2 \frac{w}{2})(w/2)^2}.$$

Since $\hat{\tilde{\phi}}(w)$ is not an entire function $\tilde{\phi}(t)$ cannot have compact support. However $\hat{\tilde{\phi}}$ is analytic in a strip about the real axis so that the coefficients $\{a_n\}$ and hence $\tilde{\phi}(t)$ have exponential decay.

In this case the multiresolution analysis (MRA) of $\tilde{\phi}$ is identical to that of ϕ since $\tilde{\phi} \in V_0$ and $\{\tilde{\phi}(t - n)\}$, being the biorthogonal sequence, is the dual Riesz basis. The mother wavelets defined by (7.20) lead to a pair of Riesz bases of $L^2(\mathbb{R})$. This and many other biorthogonal bases associated with B-splines are found in the book by C. Chui [C].

A second approach due to Cohen, Daubechies and Feauveau [C-D-F] makes possible the construction of a $\tilde{\phi}$ with compact support (but not in V_0). We illustrate their procedure again with the hat function. The function $m_0(w)$ in this case in (7.19) is given by

$$m_0(w) = \cos^2 \frac{w}{2}.$$

We try to find an $\tilde{m}_0(w)$ such that the biorthogonality condition expressed in terms of m_0 and $\tilde{m}_0$

$$m_0(w)\overline{\tilde{m}_0(w)} + m_0(w + \pi)\overline{\tilde{m}_0(w + \pi)} = 1 \tag{7.22}$$

is satisfied.

The simplest choice is $\widetilde{m_0}(w) = 1$. This does not lead to a biorthogonal function in $L^2(\mathbb{R})$ but rather the delta distribution. In fact this choice gives another approach to the sampling theorems in Chapter 9.

Another choice is

$$\tilde{m}_0(w) = \cos^2 \frac{w}{2} \left(1 + 2\sin^2 \frac{w}{2}\right)$$

for which (7.22) becomes

$$\cos^4 \frac{w}{2} \left(1 + 2\sin^2 \frac{w}{2}\right) + \sin^2 \frac{w}{2} \left(1 + 2\cos^2 \frac{w}{2}\right) = 1.$$

This is a special case of an expression in [C-D-F] for B-splines of order $2n - 1$ for which

$$m_0(w) = [\cos \frac{w}{2}]^{2n}.$$

It is

$$\tilde{m}_0(w) = \left[\cos \frac{w}{2}\right]^{2\tilde{n}} \sum_{m=0}^{n+\tilde{n}-1} \binom{n + \tilde{n} + m - 1}{m} \left[\sin^2 \frac{w}{2}\right]^m,$$

where $\tilde{n} \geq n$. Each $\tilde{m}_0$ is a trigonometric polynomial, and thus the dilation equation (7.19) has a finite number of terms. Thus by the argument in Chapter 5, the corresponding $\tilde{\phi}$ has compact support. It is shown in [C-D-F] that $\tilde{\phi}$ becomes progressively smoother as $\tilde{n}$ increases.

There is no *a priori* reason that the mother wavelets given in (7.20) generate a Riesz basis of $L^2(\mathbb{R})$. However in this special case they do [C-D-F]. Then one has the representation

$$f(x) = \sum < f, \psi_{kj} > \tilde{\psi}_{kj}(x),$$

where the $\{\psi_{kj}\}$ have small support, and the $\tilde{\psi}_{kj}$ are smooth but have greater support.

However, particularly in the case of splines the representation with $\tilde{\psi}$ and ψ interchanged may also be useful. For the splines of order 1 we have, for $f \in V_m$, the scaling function series

$$f(x) = \sum_n \langle f, \tilde{\phi}_{mn} \rangle \phi_{mn}(x).$$

However we also have the sampling theorem

$$f(x) = \sum_n f(2^{-m}n)\phi_{mn}(x)$$

in this case. Thus a simple quadrature formula

$$\langle f, \tilde{\phi}_{mn} \rangle = f(2^{-m}n)$$

is possible.

7.10   Problems

1. Find the eigenvalues and eigenfunctions of the S-L problems
   (a) $y'' = \lambda y$, $y(0) = y'(\pi) = 0$
   (b) $y'' = \lambda y$, $y(0) + y'(0) = 0$, $y(\pi) = 0$.

2. Find the Green's function $g(t, s)$ for the S-L problems in Problem 1. This involves finding a general solution to $y'' = \delta(t - s)$ and then using the boundary conditions.

3. Let $q(t,s) = \begin{cases} t(1 - s), & t \leq s \\ s(1 - t), & s < t \end{cases}$ be the kernel of an integral operator on $(0,1)$. Approximate $q(t, s)$ by the first several terms of its double Fourier series $\sum_{m=-1}^{1} \sum_{k=-1}^{1} q_{mk} e^{2\pi imt} e^{-2\pi iks}$ and find the eigenvalues of the resulting matrix $[q_{ma}]$.

4. Observe that $q(t, s)$ in Problem 3 is the Green's function of a S-L problem. Use this fact to find the exact eigenvalues.

5. Show directly that the differential operator of the Legendre polynomials

$$Py = (1 - x^2)D^2y - 2xDy$$

commutes with multiplication by the characteristic function of $[-1, 1]$.

6. Find the periodized version of the Haar wavelets and explain why it and not the Rademacher system is complete in $L^2(0, 1)$.

7. Let $h(x)$ be the characteristic function of $[-1, 1]$, with associated (ϵ, ϵ') bell $b_{\alpha,\beta}$ (7.15). Let $a_k = k$, $\epsilon_k = 1/4$. Find the associated local sine basis $v_{k,j}$.

8. Expand the function $e^{-|t|}$ in terms of the "unsatisfactory" local basis

$$w_{k,n}(t) = \frac{1}{\sqrt{2\pi}} e^{ikt} \chi(t - 2\pi n).$$

8

Pointwise Convergence of
Wavelet Expansions

Much of classical approximation theory is based on delta sequences, i.e., sequences of functions that converge to the delta distribution. For example, the Dirichlet and Fejer kernels of Fourier series and Fourier transforms, the Christoffel-Darboux formulae for orthogonal polynomials, the kernel function in statistics are all examples of delta sequences. The rate of convergence of certain types of these sequences in Sobolev spaces has recently been studied by Canuto and Quarteroni [C-Q] and Schomburg [Sc]. This rate is important for estimation in ill posed problems, in statistics [W-B], in numerical analysis, and for sampling theorems [N-W]. Canuto and Quarteroni studied delta sequences based on orthogonal polynomials while Schomburg considered kernel type sequences.

In this chapter, we consider the approximation of the delta distribution by certain partial sums of wavelet expansions. We assume that $\phi \in S_r$ and that $q_m(x, y)$ is the reproducing kernel of V_m given in Chapter 3. These reproducing kernels will constitute the delta sequence we shall be interested in.

We shall show them to be a quasi-positive delta sequence on $(-\infty, \infty)$ and then find the rate of convergence in terms of the Sobolev norms

$$\|f\|^2_{-\alpha} = \int_{-\infty}^{\infty} |\hat{f}(\omega)|^2 (\omega^2 + 1)^{-\alpha} d\omega, \qquad \alpha > \frac{1}{2}.$$

Since $\hat{\delta}(\omega) \equiv 1$ this is finite for $f = \delta$ when $\alpha > \frac{1}{2}$ but not when $\alpha \leq \frac{1}{2}$.

These delta sequences have a number of properties not shared by those arising from other orthogonal systems. These are to some extent a consequence of the following properties of the reproducing kernel of V_0

$q(x, y)$ [M, p.33]:

(i) $q(x + 1, y + 1) = q(x, y)$,

(ii) $\left| \dfrac{\partial^{\alpha+\beta}}{\partial x^\alpha \partial y^\beta} q(x, y) \right| \le C_m (1 + |x - y|)^{-m}, \quad 0 \le \alpha, \beta \le r$,

(iii) $\displaystyle\int_{-\infty}^{\infty} q(x, y) y^\alpha \, dy = x^\alpha, \quad 0 \le \alpha \le r$.

This last condition is perhaps the most surprising; it enables us to deduce that the expansion in terms of $\{\phi(x - n)\}$ of any polynomial of degree $\le r$ converges to it, as we have seen in Chapter 3.

In particular we have

$$1 = \sum_n \phi(x - n) \int \phi(y - n) dy = \sum_n \phi(x - n) \int \phi(y) dy \qquad (8.1)$$

and hence

$$\sum_n \phi(x - n) = C, \qquad x \in \mathbb{R}$$

where C is a constant $\ne 0$. From this we get

$$C \int_0^1 dx = \int_0^1 \sum \phi(x - n) dx = \int_{-\infty}^{\infty} \phi(x) dx,$$

which when substituted into (8.1) gives $1 = C^2$. Thus we may, by changing the sign if necessary, conclude that

$$1 = \int_{-\infty}^{\infty} \phi(x) dx = \sum_n \phi(x - n). \qquad (8.2)$$

8.1   Quasi-positive delta sequences

A quasi-positive delta sequence is a sequence $\{\delta_m(\cdot, y)\}$ of functions in $L^1(\mathbb{R})$ with parameter $y \in \mathbb{R}$ that satisfy:

(i)   there is a $C > 0$ such that
$$\int_{-\infty}^{\infty} |\delta_m(x, y)| dx \le C, \quad y \in \mathbb{R}, \quad m \in \mathbb{N};$$

(ii)   there is a $c > 0$ such that $\qquad\qquad\qquad\qquad$ (8.3)

$$\int_{y-c}^{y+c} \delta_m(x,y)dx \to 1$$

uniformly on compact subsets of $\mathbb{R}$, as $m \to \infty$;

(iii) for each $\gamma > 0$,

$$\sup_{|x-y|\geq\gamma} |\delta_m(x,y)| \to 0 \quad \text{as } m \to \infty.$$

It is easy to check and will follow from the next Lemma that $\delta_m(x,y) \to \delta(x-y)$ in the sense of $\mathcal{S}'$ for fixed y. An example of a quasi-positive delta sequence is the Fejer kernel,

$$F_m(x,y) = \frac{\sin^2 \frac{m+1}{2}(x-y)}{2(m+1)\pi \sin^2 \frac{x-y}{2}} \chi_{[-\pi,\pi]}(x-y),$$

while a delta sequence that is <u>not</u> quasi-positive is the Dirichlet kernel of Fourier series,

$$D_m(x,y) = \frac{\sin(m+\frac{1}{2})(x-y)}{2\pi \sin(\frac{x-y}{2})} \chi_{[-\pi,\pi]}(x-y).$$

LEMMA 8.1

Let $\{\delta_m(x,y)\}$ be a quasi-positive delta sequence; let $f \in L^1(\mathbb{R})$ be continuous on (a,b); then

$$f_m(y) = \int_{-\infty}^{\infty} \delta_m(x,y)f(x)dx \to f(y) \qquad \text{as } m \to \infty$$

uniformly on compact subsets of (a,b).

The proof of this lemma is similar to that for positive delta sequences of the form $\delta_m(x,y) = \delta_m(x-y)$ given in [K, p.33] or for bounded intervals in [Z, p.87]. We present it here in the interests of completeness.

Let $\gamma > 0$; then

$$f_m(y) = \int_{y-\gamma}^{y+\gamma} \delta_m(x,y)f(x)dx + \int_{y+\gamma}^{\infty} + \int_{-\infty}^{y-\gamma}$$

$$= f(y) \int_{y-\gamma}^{y+\gamma} \delta_m(x,y)dx \qquad\qquad (8.4)$$

$$+ \int_{y-\gamma}^{y+\gamma} \delta_m(x,y)(f(x)-f(y))dx + \left\{ \int_{y+\gamma}^{\infty} + \int_{-\infty}^{y-\gamma} \right\}$$

$$= \quad I_1 \quad + \quad I_2 \quad + \quad I_3.$$

Now let $[\alpha,\beta] \subset (a,b)$ be a closed subinterval containing the compact subset, and let $y \in [\alpha,\beta]$. We choose γ in (8.4) so that $0 < \gamma < c$,

$\beta + \gamma < b$, and $\alpha - \gamma > a$. Then for any $0 < \epsilon < 1$, we restrict f further so that

$$|f(x) - f(y)| < \epsilon \quad \text{for} \quad x \in [a, b] \quad \text{and} \quad |x - y| < \gamma.$$

From this we deduce that

$$|I_2| \le \epsilon \int_{y-\gamma}^{y+\gamma} |\delta_m(x, y)| dx \tag{8.5}$$

and that

$$|I_3| \le \sup_{\gamma \le |x-y|} |\delta_m(x, y)| \, \|f\|_1 < \epsilon \|f\|_1, \quad m \ge M_1, \tag{8.6}$$

where we have chosen M_1 so large that $\sup_{\gamma \le |x-y|} |\delta_m(x, y)| < \epsilon$ for $m \ge M_1$. We then choose $M_2 \ge M_1$ so large that

$$\left| 1 - \int_{y-\gamma}^{y+\gamma} \delta_m(x, y) dx \right| < \epsilon, \quad m \ge M_2. \tag{8.7}$$

This is possible since by (8.3) (iii)

$$\int_{y+\gamma}^{y+c} \delta_m(x, y) dx \to 0$$

uniformly on $\mathbb{R}$ as $m \to \infty$. We now combine (8.5), (8.6), and (8.7) to get

$$|f(y) - f_m(y)| \le |f(y) - I_1| + |I_2| + |I_3|$$
$$\le |f(y)| \left| 1 - \int_{y-\gamma}^{y+\gamma} \delta_m(x, y) dx \right| + \epsilon \int_{-\infty}^{\infty} |\delta_m(x, y)| dx + \epsilon \|f\|_1$$
$$\le \sup_{y \in [\alpha, \beta]} |f(y)| \epsilon + \epsilon C + \epsilon \|f\|_1$$

for $m \ge M_2$, which gives us the desired uniform convergence. $\square$

We now show that the reproducing kernels $q_m(x, y)$ constitute a quasi-positive delta sequence [W3].

LEMMA 8.2
Let $q_m(x, y)$ be the reproducing kernel of V_m, $\phi \in S_r$; then $\{q_m(x, y)\}$
satisfies (8.3) (i), (ii), and (iii).

PROOF   The inequality (8.3) (i) follows from the fact that

$$\int_{\infty}^{\infty} |q_m(x, y)| dx = \int_{-\infty}^{\infty} 2^m |q(2^m x, 2^m y)| dx$$

$$= \int_{-\infty}^{\infty} |q(x, 2^m y)| dx \le C_2 \int_{-\infty}^{\infty} (1 + |x - 2^m y|)^{-2} dx = C.$$

The second condition (ii) is obtained by a change of scale; let $c > 0$,
then

$$\int_{y-c}^{y+c} q_m(x, y) dx = \int_{2^m(y-c)}^{2^m(y+c)} q(x, 2^m y) dx$$

$$= \int_{t-2^m c}^{t+2^m c} q(x, t) dx$$

$$= 1 - \int_{t+2^m c}^{\infty} - \int_{-\infty}^{t-2^m c} = 1 - I_1 - I_2.$$

I_1 satisfies

$$|I_1| \le C_2 \int_{t+2^m c}^{\infty} \frac{1}{1 + (t - x)^2} dx = C_2 \int_{2^m c}^{\infty} \frac{1}{1 + x^2} dx \to 0$$

as $m \to \infty$ and the same for I_2. The third condition also follows from
the same inequality for any $\gamma > 0$.   □

COROLLARY 8.3
Let $f \in L^1 \cap L^2(\mathbb{R})$, continuous on (a, b); let f_m be the projection of f
onto V_m; then

$$f_m \to f \qquad \text{as} \quad m \to \infty$$

uniformly on compact subsets of (a, b).

This follows from the two lemmas since the projection is given by

$$f_m(y) = \int_{-\infty}^{\infty} q_m(x, y) f(y) dy.$$

8.2   Local convergence of distribution expansions

The global convergence of the expansions of distributions in (S_r') discussed in Chapter 5 is important for theoretical purposes. However, for

computational purposes, it is desirable to have some sort of local convergence. In the case of distributions we should like to have it converge to the value of the distribution at a point [Lo]. This is a weaker version of continuity of a function at a point.

DEFINITION 8.1
Let $f \in S'$; then f is said to have a value γ of order r at x_0 if there exists a continuous function $F(x)$ of polynomial growth such that $D^r F = f$ in some neighborhood of x_0 and

$$\lim_{x \to x_o} \frac{F(x)}{(x - x_0)^r} = \frac{\gamma}{r!}.$$

EXAMPLES The δ distribution has values 0 everywhere except at 0; the function $\sin \frac{1}{x}$, considered as an element of S', has value 0 at 0.

LEMMA 8.4 [W3]
Let $q_m(x, t)$ be the reproducing kernel of V_m, with $\phi \in S_r$, let $0 \le \alpha \le r$; then

$$K_m(x, t) := \frac{(x - t)^\alpha}{\alpha!} \frac{\partial^\alpha}{\partial t^\alpha} q_m(x, t)$$

is a quasi-positive delta sequence on $\mathbb{R}$.

PROOF We must show that the three conditions given in (8.3) hold. To prove (i) we observe first that since $|\phi^{(\alpha)}(x)| \le C_{\alpha k}(1 + |x|)^{-k}$,

$$\int |K_0(x, t)| dt \le \sum_{k=0}^{\alpha} \binom{\alpha}{k} \sum_n |\phi(x - n)| \, |x - n|^k \int |\phi^{(\alpha)}(t)| t^{\alpha - k} dt \le C,$$

a constant. The same holds for $K_m(x, t)$ since it is given by

$$K_m(x, t) = \frac{[2^m(x - t)]^\alpha}{\alpha!} \frac{\partial^\alpha}{\partial (2^m t)^\alpha} 2^m q_\alpha(2^m x, 2^m t) = 2^m K_0(2^m x, 2^m t).$$

To prove (ii) we observe first that

$$\int_{-\infty}^{\infty} K_m(x, y) dy = \int_{-\infty}^{\infty} q_m(x, y) dy = 1$$

by integration by parts. Thus it remains to be shown that for some $c > 0$,

$$\left\{ \int_{x+c}^{\infty} + \int_{-\infty}^{x-c} \right\} K_m(x, y) dy \to 0$$

as $m \to 0$ uniformly for x in a bounded set. This is shown by the same sorts of tricks as in the proof of (i),

$$\int_{x+c}^{\infty} K_m(x,y)dy = \int_{2^m(x+c)}^{\infty} K_0(2^m x, z)dz$$

$$= \sum_n \sum_{k=0}^{\alpha} (2^m x - n)^k \binom{\alpha}{k} \phi(2^m x - n)$$

$$\cdot \int_{2^m(x+c)}^{\infty} \phi^{(\alpha)}(z-n)(z-n)^{\alpha-k}dz.$$

This last integral satisfies

$$\left| \int_{2^m(x+c)}^{\infty} \phi^{(\alpha)}(z-n)(z-n)^{\alpha-k}dz \right| \leq \int_{2^m(x+c)-n}^{\infty} \frac{C_{\alpha j}}{(1+|z|)^j} |z|^{\alpha-k}dz$$

$$\leq \frac{C_{\alpha j}}{(1+|2^m(x+c)-n|)^{j-\alpha+k-2}} \int_{-\infty}^{\infty} \frac{1}{(1+|z|)^2}dz, \ j \geq \alpha - k + 2.$$

Hence

$$\left| \int_{x+c}^{\infty} K_m(x,y)dy \right| \leq C \sum_n \frac{1}{(|2^m x - n| + 1)^p} \frac{1}{(1+|2^m(x+c)-n|)^p}$$

for all $p \geq 1$. Since

$$\frac{1}{1+|x-a|}\frac{1}{1+|x-b|} \leq \frac{1}{1+|a-b|},$$

it follows that for $p > 2$

$$\left| \int_{x+c}^{\infty} K_m(x,y)dy \right|$$

$$\leq \frac{C_p}{1+2^m c} \sum_n \frac{1}{(1+|2^m x - n|)^{p-1}} \frac{1}{(1+|2^m(x+c)-n|)^{p-q}}$$

$$= \frac{C_p}{1+2^m c} h_p(2^m x, 2^m(x+c))$$

where

$$h_p(x,y) := \sum_n \frac{1}{(1+|x-n|)^{p-1}} \frac{1}{(1+|y-n|)^{p-1}}$$

is uniformly bounded in x and y. Hence (ii) holds.

For (iii) we need a bound

$$\sup_{|x-y|>\gamma} |K_m(x,y)|$$

$$\leq \sup_{|x-y|>\gamma} 2^m \sum_n |\phi(2^m x - n)|(x-y)^\alpha 2^{\alpha m}\phi^{(\alpha)}(2^m y - n)|$$

$$= \sup_{|z-w|>2^m\gamma} 2^m \sum_n |\phi(z-n)|\,|z-w|^\alpha\,|\phi^{(\alpha)}(w-n)|$$

$$\leq \sup_{|z-w|\geq 2^m\gamma} 2^m \frac{C}{(1+|z-w|)^2} h_p(z,w) \leq \frac{C\|h_p\|_\infty 2^m}{(1+2^m\gamma)^2}$$

given by the same argument as for (ii). □

THEOREM 8.5 [W3]
Let $f \in S'_{r-1}$ and have a value γ of order $\alpha \leq r$ at $x = x_0$, then the function f_m given by $f_m(x) = (f, q_m(x, \cdot))$ satisfies

$$f_m(x_0) \to \gamma \qquad \text{as } m \to \infty.$$

PROOF Each $f \in S'_{r-1}$ is a derivative of order $\beta \leq r$ of a continuous function G of polynomial growth. We may assume, by adding a polynomial of degree $< \beta$ if necessary and by increasing α if necessary that $G = F$ and $\alpha = \beta$. Then

$$f_m(x) = \int_{-\infty}^{\infty} (-1)^\alpha \partial_y^\alpha q_m(x,y) F(y) dy$$

$$= \int_{-\infty}^{\infty} \frac{(x-y)^\alpha}{\alpha!} \partial_y^\alpha q_m(x,y) \frac{F(y)\alpha!}{(y-x)^\alpha} dy$$

$$= \int_{x-A}^{x+A} + \int_{x+A}^{\infty} + \int_{-\infty}^{x-A} .$$

By a repetition of the argument in the proof in the Lemma 8.4 of (ii) we see that

$$\int_{x+A}^{\infty} + \int_{-\infty}^{x-A} \to 0 \qquad \text{as } m \to \infty.$$

Hence we may express f_m as

$$f_m(x) = \int_{-\infty}^{\infty} K_m(x,y) F_A(x,y) dy$$

where $F_A(x,y)$ is continuous for all y except for $y = x \pm A$ and has compact support. Clearly, F_A is bounded as well. Hence

$$f_m(x_0) \to F_A(x_0, x_0) \qquad \text{as } m \to \infty$$

by the property of quasi-positive delta sequences. Since $F_A(x_0, x_0) = \gamma$, the theorem is proved. □

8.3   Convergence almost everywhere

If f is not continuous but is merely in $L^2(\mathbb{R})$, we cannot expect f_m to converge to f pointwise everywhere since f can be completely arbitrary on a set of measure zero. However for Fourier series we have, by Carleson's theorem [E, p.162] convergence almost everywhere to $f(x)$ for $f \in L^2(0, 2\pi)$. We would expect the same thing to hold for wavelets since their convergence properties are usually stronger than Fourier series. This was shown to be true by S. Kelly [Ke] for $f \in L^p(\mathbb{R})$, $1 \le p \le \infty$. This is again stronger than the result for Fourier series since there exist functions in $L^1(0, 2\pi)$ whose Fourier series diverge everywhere [Z, p.310].

THEOREM 8.6
Let $f \in L^p(\mathbb{R})$, $1 \le p \le \infty$, and let x be in the Lebesgue set of f; then

$$f_m(x) \to f(x) \qquad \text{as } m \to \infty.$$

In particular $f_m \to f$ almost everywhere.

The Lebesgue set is composed of all x such that

$$\frac{1}{r} \int_{-r}^{r} (f(x, t) - f(x)) dt \to 0.$$

This holds almost everywhere on $\mathbb{R}$ if f is locally integrable [S-W, p.12]. The proof of the theory involves inequalities similar to those in Section 8.1 and will be omitted. A complete version may be found in [Ke, p.21].

8.4   Rate of convergence of the delta sequence

In order to find a rate of convergence of the wavelet expansion, we need to determine the rate at which $q_m(x, y)$ converges to $\delta(x - y)$. This rate will be in terms of the Sobolev norms mentioned at the start of the chapter.

The hypothesis will require some properties of the Zak transform [H-W] of the function ϕ,

$$Z\phi(t, \omega) := \sum_k e^{-i\omega k} \phi(t - k). \qquad (8.8)$$

Since $\phi \in \mathcal{S}_r$, this series will converge to a function which is periodic and C^∞ in ω. By (8.2), $Z\phi(t,0) = 1$ and hence $e^{i\omega t}(Z\phi)(t,\omega) = 1 + 0(|\omega|)$. Also by (8.2) $\hat{\phi}(0) = 1$, and similarly $\hat{\phi}(\omega) = 1 + 0(|\omega|)$.

In some cases, such as the Meyer wavelet case, $\hat{\phi}(\omega) \equiv 1$ in a neighborhood of $\omega = 0$, and $\hat{\phi}(\omega) = 1 + 0(|\omega|^\lambda)$ for λ arbitrarily large. This is true also for the Zak transform. Therefore, the following definition is always satisfied for some $\lambda \geq 1$.

DEFINITION 8.2
Let $\phi \in \mathcal{S}_r$ be a scaling function; then ϕ satisfies __property Z_λ__ if

$\quad$ (i) $\quad \hat{\phi}(\omega) = 1 + 0(|\omega|^\lambda) \quad$ as $\omega \to 0$

$\quad$ (ii) $\quad (Z\phi)(t,\omega) = e^{-i\omega t}(1 + 0(|\omega|^\lambda))$ uniformly as $\omega \to 0$.

This property Z_λ is related to the derivatives of $\hat{\phi}(w)$ at $w = 0$. For $\phi \in \mathcal{S}_r$, $\hat{\phi} \in C^\infty$ and hence if $\hat{\phi}^{(k)}(0) = 0$, $k = 1, 2, \cdots, \lambda - 1$, then

$$\hat{\phi}(w) = 1 + O(|w|^\lambda) \qquad \text{as } |w| \to 0.$$

For the Zak transform we have

$$\frac{d^k}{dw^k} e^{iwt}(Z\phi)(t,w)_{\omega=0}$$
$$= \sum_n (i(t-n))^k \phi(t-n).$$

Since we have the property that

$$\int x^k q(x,t)dx = t^k, \qquad k = 0, 1, \cdots, r$$

(Chapter 3) for $\phi \in \mathcal{S}_r$, it follows that

$$t^k = \int x^k \sum_n \phi(x-n)\phi(t-n)dx$$

$$= \sum_n \int \sum_{j=0}^{k} \binom{k}{j}(x-n)^j (n)^{k-j}\phi(x-n)dx \, \phi(t-n) \qquad (8.9)$$

$$= \sum_n \sum_{j=0}^{k} \binom{k}{j} n^{k-j} \int x^j \phi(x)dx \, \phi(t-n) = \sum_n n^k \phi(t-n)$$

provided $\hat{\phi}^{(j)}(0) = 0$, $j = 1, 2, \cdots, k$. Moreover, $\sum_n \phi(t-n) = 1$ and therefore

$$\sum_n (t^j - n^j)\phi(t-n) = 0, \qquad j = 0, 1, 2, \cdots, k.$$

From this it is a simple induction step to show that

$$\sum_n (t-n)^k \phi(t-n) = 0, \qquad k = 1, 2, \cdots \min(r, \gamma - 1). \qquad (8.10)$$

Hence the Zak transform satisfies

$$e^{iwt}(Z\phi)(t, w) = 1 + O(|w|^\gamma)$$

provided $\hat{\phi}^{(k)}(0) = 0$, $k = 1, 2, \cdots, \min(r, \gamma - 1)$, and we can conclude that $\phi \in Z_\lambda$ if $r \geq \lambda - 1$ and $\hat{\phi}^{(k)}(0) = 0$, $k = 1, 2, \cdots, \lambda - 1$.

Before we state the result we make two more observations. Since $\phi \in L^1(\mathbb{R})$, $|\hat{\phi}(w)| \leq \hat{\phi}(0) = 1$, and

$$\int_0^{2\pi} |Z\phi(t, \omega)|^2 d\omega = 2\pi \sum_k |\phi(t - k)|^2$$

$$\leq 2\pi \sum_k \frac{C_1^2}{(1 + |t - k|)^2} \leq C \qquad (8.11)$$

where C is a constant.

LEMMA 8.7
Let $\phi \in S_r$ be a scaling function that satisfies property Z_λ for some $\lambda > 0$; let $q_m(x, y)$ be the reproducing kernel of V_m; then

$$\|q_m(\cdot, y) - \delta(\cdot - y)\|_{-\alpha} = O(2^{-m\lambda}) \quad \text{uniformly for } y \in \mathbb{R}, \qquad (8.12)$$

where $\| \quad \|_{-\alpha}$ is the Sobolev norm and $\alpha > \lambda + \frac{1}{2}$.

PROOF The proof is the same as that in [W4] with a few simplifications.

Let $\epsilon(x, y) = q(x, y) - \delta(x - y)$ and denote by ϵ_m

$$\epsilon_m(x, y) = 2^m \epsilon(2^m x, 2^m y).$$

The left side of (8.12) can be expressed as

$$\|\epsilon_m(\cdot, y)\|_{-\alpha}^2 = \int_{\mathbb{R}} (1 + \omega^2)^{-\alpha} |\hat{\epsilon}_m(\omega, y)|^2 d\omega$$

$$= 2^m \int_{\mathbb{R}} (1 + 2^{2m}\xi^2)^{-\alpha} |\hat{\epsilon}(\xi, 2^m y)|^2 d\xi \qquad (8.13)$$

$$= 2^m \left\{ \int_{|\xi| \leq 1} + \int_{|\xi| > 1} \right\} = I_1 + I_2.$$

Since the Fourier transform of $q(x, y)$ is given by

$$\hat{q}(\omega, y) = \sum_k e^{-i\omega k} \hat{\phi}(\omega)\phi(y - k)$$

$$= \hat{\phi}(\omega)Z\phi(y, \omega), \tag{8.14}$$

it follows from condition Z_λ that

$$\hat{q}(\omega, y) = e^{-i\omega y}(1 + 0(|\omega|^\lambda)),$$

and

$$\hat{\epsilon}(\omega, y) = e^{-i\omega y}0(|\omega|^\lambda).$$

Hence the first integral in (8.13) satisfies

$$|I_1| \leq 2^{m+1} \int_0^1 (1 + 2^{2m}\xi^2)^{-\alpha} C_1 |\xi|^{2\lambda} d\xi$$

$$\leq 2^{-2m\lambda} C_2 \int_0^{2^m} (1 + \omega^2)^{-\alpha} \omega^{2\lambda} d\omega = 0(2^{-2m\lambda}). \tag{8.15}$$

The second integral satisfies

$$|I_2| \leq 2^m \int_{|\xi|>1} (1 + 2^{2m}\xi^2)^{-\alpha} |\hat{q}(\xi, 2^m y) - e^{-i\xi 2^m y}|^2 d\xi$$

$$\leq 2^{m(1-2\alpha)+1} \int_{|\xi|>1} |\xi|^{-2\alpha} \left(|\hat{q}(\xi, 2^m y)|^2 + 1\right) d\xi.$$

Since $|\hat{\phi}(\omega)| \leq \|\phi\|_1$, we have by (8.14)

$$|I_2| \leq 2^{m(1-2\alpha)+1} \int_{|\xi|>1} |\xi|^{-2\alpha} \left(|\hat{\phi}(\xi)|^2 |Z\phi(2^m y, \xi)|^2 + 1\right) d\xi$$

$$\leq 2^{m(1-2\alpha)+2} \|\phi\|_1 \sum_{k=0}^{\infty} \int_0^{2\pi} |Z\phi(2^m y, \xi + 1)|^2 |\xi + 1 + 2\pi k|^{-2\alpha} d\xi$$

$$+ \frac{2^{m(1-2\alpha)+2}}{2\alpha - 1}$$

$$\leq 2^{m(1-2\alpha)+2} \left\{ \sum_{k=0}^{\infty} |2\pi k + 1|^{-2\alpha} C + \frac{1}{2\alpha - 1} \right\} \tag{8.16}$$

$$= O(2^{-m(2\alpha-1)})$$

where C is a constant. By using (8.15) and (8.16) in (8.13) and observing that $2^{-m(2\alpha-1)} < 2^{-2\lambda m}$, we obtain the required inequality. $\square$

COROLLARY 8.8
Let $f \in H^\alpha$, let ϕ be as in Lemma 8.7, $\alpha > \lambda + \frac{1}{2}$; then the projection f_m of f onto V_m satisfies

$$\|f - f_m\|_\infty = O(2^{-m\lambda}).$$

This follows from the Schwarz inequality for Sobolev spaces. If $f \in H^\alpha$ and $g \in H^{-\alpha}$, then since $H^{-\alpha}$ is the dual space to H^α,

$$|\langle g, f \rangle|^2 = \left| \int \hat{g}(\omega)\overline{\hat{f}(\omega)}d\omega \right|^2 \le \|g\|^2_{-\alpha}\|f\|^2_\alpha,$$

and hence

$$\begin{aligned} |f(y) - f_m(y)| &= |\langle f, \delta(\cdot - y) - q_m(\cdot, y)\rangle 1 \\ &\le \|f\|_\alpha \|\delta(\cdot - y) - q_m(\cdot, y)\|_{-\alpha} = O(2^{-m\lambda}). \end{aligned}$$

8.5   Other partial sums of the wavelet expansion

The results of the last section may be given in terms of a partial sum of a wavelet expansion since

$$\begin{aligned} f_m(x) &= \sum_{n=-\infty}^{\infty} a_{mn} 2^{m/2}\phi(2^m x - n) \\ &= \sum_{k=-\infty}^{m} \sum_{n=-\infty}^{\infty} b_{kn} 2^{k/2}\psi(2^k x - n), \end{aligned} \qquad (8.17)$$

where a_{mn} are the scaling coefficients and b_{kn} are the wavelet coefficients. In practice all of the infinite sums in (8.17) must be truncated, and partial sums of the form

$$f_{m,p,r,s}(x) = \sum_{k=-p}^{m} \sum_{n=-r}^{s} b_{kn} 2^{k/2}\psi(2^k x - n). \qquad (8.18)$$

must be considered.
We first consider the outer sums. Since

$$f_m(x) - f_{-p-1}(x) = \sum_{k=-p}^{m} \sum_{n=-\infty}^{\infty} b_{kn} 2^{k/2}\psi(2^k x - n),$$

we need merely find bounds on $f_m(x)$ as $m \to -\infty$. We already know, from the multiresolution analysis, that

$$\|f_m\|^2 \to 0 \qquad \text{as} \quad m \to -\infty,$$

but we don't know anything about the uniform convergence.

We again use the Sobolev norm applied to the reproducing kernel

$$
\begin{aligned}
\|q_m(\cdot,y)\|^2_{-\alpha} &= 2^m \int (1 + 2^{2m}\xi^2)^{-\alpha} |\hat{q}(\xi, 2^m y)|^2 d\xi \\
&= 2^m \int (1 + 2^{2m}\xi^2)^{-\alpha} |\hat{\phi}(\xi)|^2 |Z\phi(2^m y, \xi)|^2 d\xi.
\end{aligned}
\tag{8.19}
$$

Since the Zak transform satisfies

$$|Z\phi(t,\xi)| \le \sum_n |\phi(t-n)|$$

and hence is uniformly bounded, (8.19) is dominated by

$$\|q_m(\cdot,y)\|^2_{-\alpha} \le 2^m \int_{-\infty}^{\infty} |\hat{\phi}(\xi)|^2 d\xi = (2\pi)2^m.$$

Thus for all $\alpha \ge 0$, we have the desired condition.

COROLLARY 8.9
Let f, ϕ be as in Corollary 8.4; let $f_{m,p,r,s}(x)$ be given by (8.18); then

$$f_{m,p,r,s}(x) \to f(x) \qquad \text{as} \quad m, p, r, s \to \infty$$

uniformly on compact subsets of $\mathbb{R}$.

The remaining calculations are routine and will be omitted.

REMARK The two convergence properties considered in this chapter, uniform convergence of the wavelet expansions for continuous functions and more rapid convergence for smoother functions, are not shared by the usual classical orthogonal systems. For example, the partial sums of the Fourier series of a continuous function do not necessarily converge. In fact there are continuous functions whose Fourier series diverge at a dense set of points [Z, p.298]. To overcome this lack of convergence, summability methods are frequently used instead. These give convergence for continuous functions but do not converge more rapidly for smooth functions [Z, p.122].

8.6   Gibbs' phenomenon

As we saw in Chapter 4, the Fourier series of a function with a jump
discontinuity exhibits Gibbs' phenomenon. That is, the partial sums
overshoot the function near the discontinuity, and this overshoot con-
tinues no matter how many terms are taken in the partial sum. More
precisely, if $f(x)$ is a piecewise Lipschitz continuous function with a positive jump discontinuity at $x = a$, then there is a sequence $x_n \to a^+$
such that $\lim_{n \to \infty} s_n(x_n) > f(a^+)$. Here $s_n(x)$ is the sequence of partial
sums of the Fourier series.

Gibbs' phenomenon does not occur if the partial sums are replaced
by the Cesàro means $\sigma_n(x)$, i.e., the average of the partial sums [Z,
p.110]. Since the wavelet expansions have convergence properties similar
to Cesàro summability, we might expect them not to exhibit Gibbs'
phenomenon. However, S. Kelly has shown [Ke] that certain wavelet
expansions do exhibit this overshoot phenomenon at the origin.

LEMMA 8.10
The wavelet expansion for $\phi \in S_r$ with reproducing kernel $g(x,t)$ exhibits
Gibbs' phenomenon at 0 if there is an $a > 0$ such that

$$\int_0^\infty g(a,t)dt > 1 \qquad (8.20)$$

(and/or an $a < 0$ such that $\int_{-\infty}^0 g(a,t)dt < 0$).

PROOF The proof involves the projection onto V_m of one of the Haar
functions (Chapter 1). We take h to be the function

$$h(t) = \chi_{[0,1)}(t) - \chi_{[0,1)}(t+1)$$

and show that there is a sequence $\{x_n\}$ such that

$$\lim_{m \to \infty} h_m(x_m) > 1$$

if (8.20) is satisfied. Here h_m is the projection onto V_m, and

$$h_m(x) = \int q_m(x,t)h(t)dt$$

$$= \int_0^1 q_m(x,t) - q_m(x,-t)dt$$

$$= \int_0^{2^m} [q(2^m x,t) - q(2^m x,-t)]dt.$$

Let $x = 2^{-m}a$ where a is the value in (8.20). Then

$$h_m(2^{-m}a) = \int_0^{2^m} [q(a,t) - q(a,-t)]dt \to \int_0^\infty q(a,t) - q(a,t)dx$$

$$= 2\int_0^\infty q(a,t) - \int_{-\infty}^\infty q(a,t)dt = 2\int_0^\infty q(a,t)dt - 1 > 1.$$

This proves the first condition is sufficient to show the phenomenon for this particular function; the second is proved the same way. Any other piecewise Lipschitz continuous function with a jump discontinuity at 0 may be made continuous in a neighborhood of 0 by adding an appropriate multiple of $h(t)$. $\square$

With this basic lemma it is possible to show that expansions in terms of many of the standard wavelet expansions exhibit Gibbs' phenomenon. In particular, many of the Daubechies wavelets [Ke] satisfy this condition of the lemma for $a \simeq 1$. It was shown by Shim [Si] that the Shannon wavelets, the Franklin wavelets, and certain of the Meyer type wavelets satisfy this condition.

The Haar wavelets do not and hence do not exhibit Gibbs' phenomenon. This can also be shown directly since $q_m(x,t)$ is a <u>positive</u> delta sequence in this case [Z, p.86].

For our other prototype, the Shannon wavelets, the condition is related to that for Fourier series (Chapter 4) since (8.20) becomes

$$\int_0^\infty \frac{\sin \pi(t - x)}{\pi(t - x)}dt = \frac{1}{\pi}\int_{-\pi x}^\infty \frac{\sin s}{s}ds$$

$$= \frac{1}{\pi}\int_0^\infty \frac{\sin s}{s}ds + \frac{1}{\pi}\int_{-\pi x}^0 \frac{\sin s}{s}ds.$$

If $x = 1$, then $\frac{1}{\pi}\int_{-\pi}^0 \frac{\sin s}{s}dx > \frac{1}{2}$ and therefore the condition is satisfied ($\int_0^\infty \frac{\sin}{s}ds = \pi/2$).

In order to get around Gibbs' phenomenon, one can sometimes use other approximations to $f \in V_m$ rather than the projection. In the case of the Shannon wavelet, the function $\sigma(x) = \frac{\sin^2 \pi x/2}{(\pi x/2)^2}$ belongs to V_0 and in fact, $f * \sigma \in V_0$ for each $f \in L^2(\mathbb{R})$. It follows (with a little work [Si]) that

$$\delta_m(x,y) := 2^m \sigma(2^m(x - y))$$

is a <u>positive</u> delta sequence, and hence the function

$$f_m^1(x) = \int_{-\infty}^\infty 2^m \sigma(2^m(x - y))f(y)dy$$

is an approximation to $f(x)$ which does not show Gibbs' phneomenon.

8.7   Problems

1. For a <u>positive</u> delta sequence, the conditions are the same as
   for a quasi-positive (8.3) except that conditions (i) is replaced
   by

   $$(i)'\quad \delta_m(x,y) \geq 0\,,\quad x,y \in \mathbb{R}\,,\quad m \in \mathbb{N}$$

   $$\text{and}\quad (iv)'\quad \sup_y \int_{|x-y|>\gamma} \delta_m(x,y)dx \to 0\quad \text{as } m \to \infty$$

   $$\text{for some}\quad \gamma > 0\,.$$

   Show that a positive delta sequence is also quasi-positive.

2. Show that the Fejer kernel is a positive delta-sequence.

3. Let $f \in V_0$ for some r-regular scaling function ϕ and let $f(x) \geq 0$, $\hat{f}(2\pi k) = \delta_{ok}$. Show that

   (a) $\int\limits_{-\infty}^{\infty} f(x)dx = 1$

   (b) $\sum\limits_n f(x-n) = 1$

   (c) $\sum\limits_n f^2(x-n) > 0 \quad x \in \mathbb{R}$

   (d) $\sum\limits_n f(x-n)f(y-n) \leq C(1+|x-y|)^{-1}$.

4. Use the results from Problem 3 to show that

   $$k_m(x,y) := 2^m \sum_n f(2^m x - n)f(2^m y - n)$$

   is a positive delta sequence.

5. Show that the reproducing kernel $q_m(x,y)$ of V_m for the Haar
   wavelet is a positive delta sequence.

6. Let $k_m(x,y)$ be a positive delta sequence, let $f(x) \leq M$, $x \in \mathbb{R}$.
   Show that

   $$f_m(x) = \int k_m(x,y)f(y)dy \leq M\,.$$

7. Show directly that the expansion of $\delta(x)$, in terms of 1- regular
   wavelets, converges to zero pointwise for all $x \neq 0$.

8. Show that if the tempered distribution f is equal to a continuous function $g(x)$ in some neighborhood of x_o, then f has the value $g(x_o)$ at x_o.

9. Show that for the Franklin wavelets, ϕ satisfies property Z_2 and for the Meyer wavelets ϕ satisfies Z_λ for all $\lambda \geq 1$.

9

A Shannon Sampling Theorem in V_m

The classical Shannon sampling theorem [S] given by the formula [Sh]

$$f(t) = \sum_{n=-\infty}^{\infty} f(nT)\frac{\sin \sigma(t - nT)}{\sigma(t - nT)}, \quad T = \pi/\sigma, \tag{9.1}$$

holds for σ-band limited signals, i.e., continuous functions in $L^2(\mathbb{R})$ whose Fourier transform has support in $[-\sigma, \sigma]$. If σ is taken to be $\sigma = 2^m\pi$, and $\phi(t) = \sin \pi t/\pi t$ is the scaling function of Example 5, then (9.1) is a statement about the elements of the subspaces V_m in the multiresolution analysis.

Many generalizations of this theorem have been proposed. They usually involve transforms other than the Fourier transform [B-S-S], [H], and can be put into a reproducing kernel Hilbert space setting [N-W]. The original theorem has widespread applications in communication theory since it enables one to recover an analog signal from its sampled values. It and many others can be imbedded in wavelet theory, which gives a unified approach to the subject [W5], [W6].

There are other advantages to this wavelet approach to sampling theorems as well.

For example, the aliasing error may be given in terms of the wavelet coefficients [W5]. The "mother wavelet" $\psi(t)$ may be found in this case to be

$$\psi(t) = \frac{\sin \pi(t - \frac{1}{2}) - \sin 2\pi(t - \frac{1}{2})}{\pi(t - \frac{1}{2})}.$$

The translates of ψ, $\{\psi(t - n)\}$ are an orthonormal basis of W_0, the orthogonal complement of V_0 in V_1, which is composed of $f \in L^2(\mathbb{R})$ whose Fourier transform has support in $[-2\pi, -\pi] \cup [\pi, 2\pi]$.

In this chapter we shall turn this around and first ask if there always is a sampling theorem imbedded in any wavelet theory. That is, if we begin with any scaling function satisfying the required properties, is there a natural sampling function $S_n(t)$ that gives a sampling expansion for $f \in V_0$? Then, of course, by a change of scale, we have the same for all V_m. We shall see that there is indeed such a sampling expansion under very broad conditions. We shall first introduce another basis of the V_m spaces and use it to construct the sampling functions. This will work in some cases; in others we must extend V_m to the space T_m given in Chapter 5.

Next we consider a generalization to shifted sampling, i.e., we recover $f \in V_0$ from its values at $(a + n)$. This approach, due to Janssen, works in other cases where an unshifted sampling function doesn't exist [J].

The final sections are concerned with the construction of wavelet subspaces in which the scaling function is also a sampling function. We discuss the work of Xia and Zhang [X-Z] and its extension to band limited scaling functions [W-L]. In addition, we consider wavelet subspaces in which the scaling function serves as a sampling function but at double the rate [W7].

9.1　A Riesz basis of V_m

A Riesz basis of a separable Hilbert space H is a basis $\{f_n\}$ that is close to being orthogonal. That is, there is a bounded invertible operator that maps $\{f_n\}$ onto an orthonormal basis, [Y, p.31].

Each $f \in H$ has a representation

$$f = \Sigma a_n f_n.$$

The coefficients $a_n = \langle f, g_n \rangle$ where $\{g_n\}$ is the unique biorthogonal basis, which is also a Riesz basis, [Y, p.36].

We shall need an alternate basis of our spaces V_m different from that based on the scaling function. It will be based on the reproducing kernel (Chapter 3, section 3)

$$q(x,t) = \sum_n \phi(x - n)\phi(t - n), \qquad (9.2)$$

by taking one of the parameters equal to an integer (ϕ is assumed real).

PROPOSITION 9.1
Let $\phi(t)$ be a scaling function in S_r such that

$$\hat{\phi}^*(\omega) := \sum_n \phi(n)e^{-i\omega n} \neq 0, \quad \omega \in \mathbb{R}; \qquad (9.3)$$

then $\{q(t,n)\}$ is a Riesz basis of V_0, with biorthogonal Riesz basis $\{S_n(t)\}$.

PROOF Let T be the operator on V_0 that takes $\phi(x-n)$ into $q(x,n)$. Its Fourier transform induced operator $\hat{T}$ which maps the Fourier transform of $\phi(x-n)$ into that of $q(x,n)$, satisfies

$$\hat{T}\hat{\phi}(\omega) = \hat{q}(\omega,0) = \sum_k \hat{\phi}(\omega)e^{-i\omega k}\phi(0-k) = \overline{\hat{\phi}^*(\omega)}\hat{\phi}(\omega).$$

Since $\{\phi(k)\}$ is rapidly decreasing, $\hat{\phi}^*(\omega)$ is continuous and periodic and hence a bounded function. Hence $\hat{T}$ satisfies

$$\|\hat{T}\hat{\phi}\| \leq \|\hat{\phi}^*\|_\infty\|\hat{\phi}\| \tag{9.4}$$

and is therefore a bounded operator on $\hat{V}_0$. Since, by (9.3), $1/\hat{\phi}^*(\omega)$ is also a bounded function, $\hat{T}^{-1}$ is bounded as well and takes $\hat{q}(\omega,0)e^{-i\omega n} = \hat{q}(\omega,n)$ into $\hat{\phi}(\omega)e^{-i\omega n}$. Since the Fourier transform is an isometry, T is also a bounded invertible operator, and hence $\{q(t,n)\}$ is a Riesz basis that must have a unique biorthogonal basis $\{S_n(t)\}$ [Y, p.32]. $\square$

9.2   The sampling sequence in V_m

THEOREM 9.2
Let $f \in V_0$, let ϕ satisfy (9.3), and let $\{S_n(t)\}$ be the basis biorthogonal to $\{q(t,n)\}$; then $S_n(t) = S(t-n)$ where

$$\hat{S}(\omega) = \hat{\phi}(\omega)/\hat{\phi}^*(\omega), \qquad \omega \in \mathbb{R} \tag{9.5}$$

and

$$f(t) = \sum_n f(n)S(t-n), \qquad t \in \mathbb{R} \tag{9.6}$$

where the convergence is uniform on $\mathbb{R}$.

PROOF Since $q(t,n) = q(t-n,0)$, it follows that $\{S_0(t-n)\}$ must also be a biorthogonal sequence, which because of uniqueness must be equal to $S_n(t)$. The expression (9.6) is just the expansion of f with respect to $\{S_n(t)\}$ (since $\int f(t)q(t,n)dt = f(n)$), which converges in the sense of $L^2(\mathbb{R})$. The calculation of $S(t) = S_0(t)$ follows at once by substituting $\phi(t)$ for $f(t)$ in (9.6), and then taking the Fourier transform,

$$\hat{\phi}(\omega) = \sum_n \phi(n)e^{-i\omega n}\hat{S}(\omega) = \hat{\phi}^*(\omega)\hat{S}(\omega).$$

This is solved for $\hat{S}(\omega)$ to get (9.5).

To show that (9.6) converges uniformly we use

$$\left| f(t) - \sum_{n=-N}^{N} f(n)S(t-n) \right|$$

$$= \left| \int_{-\infty}^{\infty} q(t,s) \left[f(s) - \sum_{n=-N}^{N} f(n)S(s-n) \right] ds \right|$$

$$\leq \left\{ \int |q(t,s)|^2 ds \right\}^{1/2} \left\| f - \sum_{n=-N}^{N} f(n)S_n \right\|.$$

Since $q(t,t) = \int |q(t,s)|^2 ds = \Sigma \phi^2(t-n)$ and hence is periodic, it must be uniformly bounded. □

Several different types of errors may occur in this and other sampling theorems. They are truncation error, jitter error, round off error, and aliasing error. We shall treat only aliasing error, which usually refers to the error in the classical Shannon sampling theorem that arises when the signal is not σ- bandlimited but rather σ' bandlimited for some $\sigma' > \sigma$. The natural analogy to this is when the function $f \in V_m$ rather than V_0 for some $m > 0$. In particular such error arises when $f \in V_1$.

DEFINITION 9.1
The aliasing error for $f \in V_1$ is the difference

$$e(t) := f(t) - \sum_{n} f(n)S(t-n).$$

PROPOSITION 9.3
Let $\{\psi(t-n)\}$ be the wavelet basis of W_0; then the aliasing error for $f \in V_1$ satisfies
$$|e(t)|^2 \leq C \sum_{n} |b_n|^2, \qquad t \in \mathbb{R}$$

where C is constant, and the $b_n = \int f(t)\overline{\psi(t-n)}dt$ are the wavelet co-efficients of f.

PROOF   Let $f = f_0 + f_1$ where $f_0 \in V_0$ and $f_1 \in W_0$; then $e(t) = f_1(t) - \Sigma f_1(n)S(t-n)$ since $f_0(t)$ is equal to its sampling series.
   But we see that

$$|f_1(t)|^2 = \left| \sum_{n} b_n \psi(t-n) \right|^2 \leq \sum_{n} |b_n|^2 \sum_{n} |\psi(t-n)|^2$$

and

$$\left| \sum_n f_1(n) S(t-n) \right|^2 \le \sum_n |f_1(n)|^2 \sum_n |S(t-n)|^2.$$

Moreover by the argument in Proposition 9.1 (applied to W_0 instead of V_0) we have

$$\sum_n |f_1(n)|^2 \le C\|f_1\|^2 = C \sum_n |b_n|^2.$$

The proof is completed by observing that both $\psi(t)$ and $S(t)$ are continuous and rapidly decreasing and hence $\Sigma|\psi(t-n)|^2$ and $\Sigma|S(t-n)|^2$ are bounded for $t \in \mathbb{R}$. $\square$

9.3   Examples of sampling theorems

We return to the examples of Chapter 3 and try to construct sampling functions. For the Haar system, the reproducing kernel is $q(t,s) = \phi(t - [s])$ and hence the basis is $q(t,n) = \phi(t-n) = S_n(t)$, the sampling function. This and the Shannon case, Example 5, are the only ones among the examples of Chapter 3 for which this happens.

The Franklin wavelets, Example 2, have

$$\hat{\phi}^*(\omega) = \sum_k \phi(k) e^{-i\omega k} = \sum_k \hat{\phi}(\omega + 2\pi k) = \left(1 - \frac{2}{3}\sin^2 \frac{\omega}{2} \right)^{-\frac{1}{2}}.$$

Thus we find

$$\hat{S}(\omega) = \hat{\phi}(\omega)/\hat{\phi}^*(\omega) = \left(\frac{\sin\omega/2}{\omega/2} \right)^2$$

whose inverse Fourier transform is the original "hat function", $S(t) = (1 - |t|)\chi_{(-1,1)}(t)$.

For cubic splines, Example 4, the space V_0 is the closed linear span of translates of the third order basic spline $\theta_3(t)$. The $\{\theta_3(t - k)\}$ are not orthogonal but may be orthogonalized in the standard way. However this is unnecessary since $\hat{S}(\omega) = \hat{\theta}(\omega)/\hat{\theta}^*(\omega)$ as well as (9.5).

Since $\theta_3(t)$ has support on $[0,4]$, only its values on the integers 1, 2, 3 are needed and are given by $\theta_3(1) = \theta_3(3) = 1/6$, $\theta_3(2) = 2/3$; therefore

$$\hat{\theta}_3^*(\omega) = \frac{1}{6} e^{-i\omega} \left[1 + 4e^{-i\omega} + e^{-2i\omega} \right],$$

which has no real zeros. The inverse $\hat{\theta}_3^{*-1}$ may be found by using the

Laurent series of

$$\frac{6}{z^{-1}[1 + 4z^{-1} + z^{-2}]} = \sqrt{3}\sum_{n=0}^{\infty}(\sqrt{3} - 2)^{n+1}z^{n-1}$$

$$+ \sqrt{3}\sum_{n=1}^{\infty}(\sqrt{3} - 2)^{n-1}z^{-n-1}.$$

From this we find that

$$S_3(t) = \sqrt{3}\sum_{n=0}^{\infty}\left(\sqrt{3} - 2\right)^{n+1}\theta_3(t - n + 1)$$

$$+ \sqrt{3}\sum_{n=1}^{\infty}\left(\sqrt{3} - 2\right)^{n-1}\theta_3(t + n + 1). \tag{9.7}$$

In Example 3, the <u>Daubechies Wavelets</u>, the scaling function has support on [0,3], and, since $\phi(j) = \sqrt{2}\Sigma c_k\phi(2j - k) = \sqrt{2}\Sigma c_{2j-k}\phi(k)$, it follows that $(\phi(0),\ \phi(1),\ \phi(2),\ \phi(3))^T$ is an eigenvector of the matrix $C = \sqrt{2}[c_{2j-k}]$. Since $\phi(0) = \phi(3) = 0$, we need consider only the matrix

$$C = \begin{bmatrix} c_1 & c_0 \\ c_3 & c_2 \end{bmatrix} = \frac{1}{\nu^2 + 1}\begin{bmatrix} 1 - \nu & \nu(\nu - 1) \\ (\nu + 1)\nu & \nu + 1 \end{bmatrix}.$$

An eigenvector corresponding to eigenvalue 1, is $\phi(1) = \frac{\nu - 1}{2\nu}$, $\phi(2) = \frac{\nu + 1}{2\nu}$, which satisfies $\sum_k \phi(k) = 1$ as it must and

$$\hat{\phi}^*(\omega) = \frac{1}{2\nu}\left\{(\nu - 1)e^{-i\omega} + (\nu + 1)e^{-2i\omega}\right\},$$

which has no zeros for $\nu \neq 0$. For $\nu < 0$ we see that

$$\hat{\phi}^{*-1}(\omega) = \frac{2\nu}{\nu - 1}\sum_{n=0}^{\infty}\left(\frac{1 + \nu}{1 - \nu}\right)^n e^{-i(n-1)\omega}$$

and hence

$$S(t) = \frac{2\nu}{\nu - 1}\sum_{n=0}^{\infty}\left(\frac{1 + \nu}{1 - \nu}\right)^n \phi(t - n + 1).$$

For $\nu > 0$, the series becomes

$$S(t) = \frac{2\nu}{\nu + 1}\sum_{n=0}^{\infty}\left(\frac{1 - \nu}{1 + \nu}\right)^n \phi(t + n + 2).$$

REMARK   The examples illustrate each of the three methods that can be used to find $\hat{\phi}^*(\omega)$. The first is the definition given by (9.3). If $\phi \in \mathcal{S}_r$ for $r \geq 2$, then the periodic extension of $\hat{\phi}$ may also be used since the series

$$g(\omega) = \sum_{k=-\infty}^{\infty} \hat{\phi}(\omega + 2\pi k)$$

then converges uniformly. Its Fourier coefficients are

$$\frac{1}{2\pi} \int_0^{2\pi} \sum_k \hat{\phi}(\omega + 2\pi k) e^{-i\omega n} \, d\omega = \frac{1}{2\pi} \int_{-\infty}^{\infty} \hat{\phi}(\omega) e^{-i\omega n} \, d\omega = \phi(-n)$$

and hence

$$g(\omega) = \sum_n \phi(-n) e^{i\omega n} = \hat{\phi}^*(\omega).$$

We used this in Example 2, even though $r = 1$ in that case.

In the case of Example 7, the sampling function is again easy to find. Since

$$\hat{\phi}^2(\omega) = \int_{\pi-\omega}^{\pi+\omega} dP$$

satisfies $\hat{\phi}(\omega) = 1$ for $|\omega| \leq 2\pi/3$ and has support on $\left[-\frac{4\pi}{3}; \frac{4\pi}{3}\right]$, it follows that

$$\hat{\phi}^*(\omega) = \hat{\phi}(\omega) + \hat{\phi}(\omega - 2\pi) + \hat{\phi}(\omega + 2\pi) \quad |\omega| \leq 4\pi/3.$$

Hence $\hat{S}(\omega) = 1$ for $|\omega| \leq 2\pi/3$ and

$$\hat{S}(\omega) = \frac{\hat{\phi}(\omega)}{\hat{\phi}(\omega) + \hat{\phi}(\omega - 2\pi) + \hat{\phi}(\omega + 2\pi)}, \qquad \omega \in \mathbb{R}.$$

If only the dilation equation is known, and only a finite number of the c_k's are not zero, then the procedure of Example 3 can be used. That is, we can look for the right eigenvector of $C = \sqrt{2}[c_{2j-k}]$ corresponding to the eigenvalue 1. Since $\sqrt{2}\Sigma c_{2j} = \sqrt{2}\Sigma c_{2j-1} = 1$ in most examples [D2, p.936], there is always a left eigenvector in these cases.

9.4   The sampling sequence in $\mathbf{T}_m$

In some cases $\hat{\phi}^*(\omega)$ has real zeros (see quadratic splines below). In these cases we can find a sampling sequence in T_m, the multiresolution subspace of $\mathcal{S}'_r$ in Chapter 5.

We begin with the same definition of the sampling function $S_n(t)$ in (9.5), but without the added condition that $\hat{\phi}^*(\omega) \neq 0$. Rather we require only that its zeros be isolated and of finite multiplicity $< r$ in $[-\pi, \pi]$. Then the function

$$\hat{\phi}^*(\omega) / \prod_j \sin \frac{(\omega - \alpha_j)}{2} e^{\frac{i(\omega - \alpha_j)}{2}} = \hat{\phi}^*(\omega)/f(\omega), \qquad (9.8)$$

where the α_i's are the zeros of $\hat{\phi}^*(\omega)$ (counted according to multiplicity), has no zeros and is periodic. Since $\hat{\phi}^*(\omega) \in C^\infty$, the reciprocal $f(\omega)/\hat{\phi}^*(\omega)$ of (9.8) is continuous provided the values at $\omega = \alpha_j$ are defined by continuity. The same is true for its derivatives. Hence it is in C^∞ and periodic, and its Fourier coefficients are rapidly decreasing.

The Fourier series of $e^{-i\omega/2}/\sin \omega/2$ may be found from the power series of

$$h(z) = \frac{-2i}{1 - z^2}$$

to be

$$\frac{e^{-i\omega/2}}{\sin \omega/2} \sim -2i \sum_{n=0}^{\infty} e^{i\omega n},$$

that converges in S'_r for $r > 0$. Each of the first order factors in $f^{-1}(\omega)$ therefore has a series that has bounded coefficients. If a factor is repeated k times the resulting series has coefficients $= O(|n|^k)$. The product of two such series has coefficients whose growth is the same order as the maximum order of the coefficient of each series. Hence $f^{-1}(\omega)$ has a Fourier series whose coefficients are $O(|n|^{r-1})$ and which therefore converges in S'_r.

Its product with a C^∞ function of period 2π again forms a Fourier series convergent in S'_r. Hence

$$\hat{\phi}^{*-1}(\omega) = \sum_n \alpha_n e^{-i\omega n}, \qquad \alpha_n = 0(|n|^r),$$

and the sampling function $S(t)$ as the inverse Fourier transform of $\hat{\phi}^{*-1}(\omega)\hat{\phi}(\omega)$ is given by

$$S(t) = \sum_n \alpha_n \phi(t - n). \qquad (9.9)$$

Hence $S(t) \in T_o$. Clearly $S(k) = \delta_{0k}$. To show the sampling series converges we observe that for $f \in T_0$ we have

$$\hat{f}(\omega) = \sum_n a_n \hat{\phi}(\omega) e^{-i\omega n} = \hat{a}^*(\omega)\hat{\phi}(\omega)$$

$$= \hat{a}^*(\omega)\hat{\phi}^*(\omega)\hat{S}(\omega)$$

the product of a periodic distribution with $\hat{S}(\omega)$. The inverse Fourier transform gives

$$f(t) = \sum_n b_n S(t-n)$$

with convergence in $\mathcal{S}_r'$. For $f(t) \in \mathcal{S}_r$, the convergence is also uniform, and we have

$$f(k) = \sum_n b_n S(k-n) = b_k.$$

Hence the sampling theorem (8.6) holds in T_0 under greater generality.

Example 4 also includes <u>quadratic splines</u>. Here $\theta_2(t)$ is the basic spline of order 2,

$$\hat{\theta}_2(\omega) = \left[\frac{1-e^{-i\omega}}{i\omega}\right]^3, \qquad \omega \in \mathbb{R}.$$

It is a quadratic polynomial in $(n, n+1)$, $n \in \mathbb{Z}$, has support in $[0,3]$ and belongs to $\mathcal{S}_1$. The periodic extension in $\hat{\theta}_2^*(\omega) = \frac{1}{2}e^{-i\omega}[1+e^{-i\omega}]$, which has a zero at $\omega = \pm\pi$.

Its reciprocal converges in the sense of $\mathcal{S}_1'$ and is used to calculate

$$S_2(t) = 2\sum_{n=0}^{\infty}(-1)^n \theta_2(t-n+1).$$

Since $S_2(t) \notin V_0$, this space has no sampling theorem. But $S_2(t) \in T_0$, and, hence, for $f \in V_0 \cap \mathcal{S}_1$, we have

$$f(t) := \sum_n f(n)S_n(t)$$

convergent in $\mathcal{S}_1'$ and uniformly pointwise convergent.

This setting enables us to treat sampling by means of a traditional engineering approach. A signal $f(t) \in T_0$ is sampled at the integers and an impulse train

$$f^*(t) = \sum_n f(n)\delta(t-n) \tag{9.10}$$

is formed. This series converges in the sense of $\mathcal{S}_r'$. The projection of $f^*(t)$ onto T_0 (actually a type of low-pass filter) gives

$$P_0 f^*(t) = \sum_n f(n)q(t-n,0)$$

where $q(t,u)$ is the reproducing kernel of V_0. Since $\hat{q}(\omega,0) = \hat{\phi}(\omega)\overline{\hat{\phi}^*(\omega)}$ and $\hat{S}(\omega) = \hat{\phi}(\omega)/\hat{\phi}^*(\omega)$, it follows that $\hat{q}(\omega,0) = |\hat{\phi}^*(\omega)|^2\hat{S}(\omega)$. Hence by multiplying $\hat{P}f^*(\omega)$ by $|\hat{\phi}^*(\omega)|^{-2}$ and taking the inverse Fourier transform we recover our original function from the sampling series (9.6). This again requires $\hat{\phi}^*(\omega) \neq 0$.

9.5 Shifted sampling

An alternate approach to that of Section 9.4 is to sample at points $\{a+n\}$ and then ask if a sampling function exists for such points. The natural hypothesis to use here is the shifted version of (9.3); there is an $a \in [0, 1)$ such that

$$(Z\phi)(a, \omega) = \sum_n \phi(a + n)e^{-i\omega n} \neq 0, \quad \omega \in \mathbb{R}.$$

This of course is the Zak transform of ϕ mentioned in Chapter 8. Janssen [J] showed that the B-spline wavelet of order 2 satisfies this condition for $a = \frac{1}{2}$, but, as we observed, it does not hold for $a = 0$.

In this case the sampling function $S_a \in V_0$ is given by

$$\hat{S}_a(\omega) = \frac{\hat{\phi}(\omega)}{(Z\phi)(a, \omega)} = \hat{\phi}(\omega) \sum_n \sigma_{n,a} e^{-i\omega n}. \tag{9.11}$$

The cardinal series of $f \in V_0$ is then

$$f(t) = \sum_n f(a + n)S_a(t - n). \tag{9.12}$$

The derivation and proof is similar to that in Section 9.2 for $a = 0$ and is omitted.

The question then arises, if several choices of a are available, which is the best? One possible criterion is the one that minimizes the aliasing error [J]. This is the error arising if $f \in V_1$ is approximated by the right side of (9.12). Since the error

$$e_{f,a}(t) = f(t) - f_0(t) + \sum_n (f(a + n) - f_0(a + n))S_a(t - n)$$

depends only on the difference between $f - f_0$ (f_0 is the projection on V_0) which belongs to W_0, we need only consider the $e_{f,a}$ for $f \in W_0$. We find that for $f \in W_0$,

$$\|e_{f,a}\|^2 = \|f\|^2 + \|\sum_n f(a + n)S_a(\cdot - n)\|^2$$

$$= \|f\|^2 + \sum_m \sum_n f(a + n)\overline{f(a + m)} \int_{-\infty}^{\infty} S_a(t - n)\overline{S_a(t - m)}dt$$

$$= \|f\|^2 + \sum_m \sum_n f(a + n)\overline{f(a + m)} \frac{1}{2\pi} \int_{-\infty}^{\infty} |\hat{S}_a(\omega)|^2 e^{i(m-n)\omega} d\omega$$

by Parseval's equality. But by (9.11) this last integral is

$$\frac{1}{2\pi} \int_{-\infty}^{\infty} |\hat{S}_a(\omega)|^2 e^{i(m-n)\omega} \, d\omega$$

$$= \frac{1}{2\pi} \int_{-\infty}^{\infty} |\hat{\phi}(\omega)|^2 \left| \sum_k \sigma_{k,a} e^{-i\omega k} \right|^2 e^{i(m-n)\omega} \, d\omega$$

$$= \sum_k \sigma_{k-n,a} \overline{\sigma_{k-m,a}}$$

by the orthogonality of $\{\hat{\phi}(\omega)e^{in\omega}\}$. Hence the error is

$$\|e_{f,a}\|^2 = \|f\|^2 + \sum_k \left| \sum_m \sigma_{k-m,a} f(m+a) \right|^2 \tag{9.13}$$

$$= \|f\|^2 + \frac{1}{2\pi} \int_0^{2\pi} \frac{|Zf(a,\omega)|^2}{|Z\phi(a,\omega)|^2} \, d\omega$$

since $Zf(a,\omega)$ has Fourier coefficients $f(n+a)$ and $\frac{1}{Z\phi(a,\omega)}$ has Fourier coefficients $\sigma_{n,a}$.

We now use the fact that $f \in W_0$ has a wavelet expansion as well,

$$f(t) = \sum_k \beta_k \psi(t-k).$$

The Zak transform of f then becomes

$$(Zf)(a,\omega) = \sum_k \beta_k e^{-i\omega k} (Z\psi)(a,\omega),$$

and the integral in (9.13) therefore satisfies

$$\frac{1}{2\pi} \int_0^{2\pi} \left| \frac{Zf(a,\omega)}{Z\phi(a,\omega)} \right|^2 d\omega = \frac{1}{2\pi} \int_0^{2\pi} \left| \frac{Z\psi(a,\omega)}{Z\phi(a,\omega)} \right|^2 \left| \sum_k \beta_k e^{-i\omega k} \right|^2 d\omega$$

$$\leq \sup_\omega \left| \frac{Z\psi(a,\omega)}{Z\phi(a,\omega)} \right|^2 \sum_k |\beta_k|^2.$$

Since $\sum |\beta_k|^2 = \|f\|^2$, the error satisfies

$$\|e_{f,a}\|^2 \leq \|f\|^2 \left(1 + \sup_\omega \left| \frac{Z\psi(a,\omega)}{Z\phi(a,\omega)} \right|^2 \right). \tag{9.14}$$

This leads to one of the criteria by Janssen [J] which is to choose a such that $\sup_\omega \left| \frac{Z\psi(a,\omega)}{Z\phi(a,\omega)} \right|$ is minimized. He found for the Daubechies wavelets that a value of $a \approx 0.37$ gives a value close to the minimum.

We have not remarked on the convergence of the various series considered here. For $\phi \in S_r$ this is no problem since $(Z\phi)(a,\omega) \in C^\infty$ in this case, and all coefficients are rapidly decreasing. The same conclusions hold under the weaker hypothesis [J] that ϕ is bounded and $\sum_n |\phi(t-n)|$ converges uniformly.

9.6 Oversampling with scaling functions

In the classical Shannon sampling theorem, the same sequence of functions is both orthonormal and a sampling sequence. This is not true for most wavelet subspaces in which the sampling functions and the orthonormal bases are different. However, by oversampling at double the rate the property of the Shannon wavelets can be extended to a much larger class, which includes the Meyer wavelets [W7]. This will be shown in this section.

More precisely, we shall look for $\phi(t)$ for which $\phi(2t-n)$ is a sampling function for V_0 by finding the properties it must satisfy. We then show that for the Meyer family, these properties are satisfied. Finally we show that the oversampling property together with another property characterize this family.

Accordingly, let $\phi(t) = 0\left((1+|t|)^{-1-\epsilon}\right)$ and $\hat{\phi}(\omega) = 0\left((1+|\omega|)^{-1-\epsilon}\right)$, $\epsilon > 0$, and let $\{\phi(t-n)\}$ be an orthonormal basis of V_0 such that for each $f \in V_0$,

$$f(t) = \sum_n f\left(\frac{n}{2}\right) \phi(2t-n), \qquad (9.15)$$

in the sense of $L^2(\mathbb{R})$. Then, in particular

$$\phi(t) = \sum_n \phi\left(\frac{n}{2}\right) \phi(2t-n)$$

and by taking Fourier transforms we find

$$\hat{\phi}(\omega) = \frac{1}{2} \sum_n \phi\left(\frac{n}{2}\right) e^{-i\omega n/2} \phi\left(\frac{\omega}{2}\right)$$

$$= \sum_n \hat{\phi}(\omega + 4\pi k) \hat{\phi}\left(\frac{\omega}{2}\right).$$

The last equality is obtained by finding the Fourier coefficients of the 4π periodic function

$$\hat{\phi}^*(\omega) := \sum_k \hat{\phi}(\omega + 4\pi k),$$

which are exactly $\frac{1}{2}\phi\left(\frac{n}{2}\right)$. Thus, the Fourier transform of the dilation equation is

$$\hat{\phi}(\omega) = \hat{\phi}^*(\omega)\hat{\phi}\left(\frac{\omega}{2}\right). \tag{9.16}$$

This is also sufficient for (9.15) to hold.

LEMMA 9.4
Let ϕ be a scaling function such that $\phi(t) = 0\left((1+|t|)^{-1-\epsilon}\right)$ and $\hat{\phi}(\omega) = 0\left((1+|\omega|)^{-1-\epsilon}\right)$ for $\epsilon > 0$; then (9.16) holds for ϕ if and only if (9.15) holds for all $f \in V_0$.

PROOF   In order to show that (9.15) holds we must first show the series converges in the sense of $L^2(\mathbb{R})$. Since $\{\sqrt{2}\,\phi(\omega t - n)\}$ is orthonormal, we need only show that $\{f\left(\frac{n}{2}\right)\} \in \ell^2$. Since $f \in V_0 \subseteq V_1$, it has an expansion convergent in $L^2(\mathbb{R})$ and, because of the decay property of $\phi(t)$, uniformly on bounded sets,

$$f(t) = \sum_n a_{n,1}\sqrt{2}\,\phi(2t - n).$$

Thus we have

$$f\left(\frac{k}{2}\right) = \sum_n a_{n,1}\sqrt{2}\,\phi(k - n)$$

and by taking the discrete Fourier transform, we find

$$\sum_k f\left(\frac{k}{2}\right)e^{i\omega k} = \sqrt{2}\sum_n a_{n,1}e^{i\omega n}\sum_k \phi(k)e^{i\omega k}. \tag{9.17}$$

The right hand side is the product of a bounded function and an $L^2(-\pi, \pi)$ function. Hence the left hand side of (9.17) is in $L^2(-\pi, \pi)$ and $\{f\left(\frac{k}{2}\right)\} \in \ell^2$.

To show that it converges to $f(t)$ we use its expansion in V_0, which is

$$f(t) = \sum_n a_{n,0}\phi(t - n)$$

$$= \sum_n a_{n,0}\sum_j \phi\left(\frac{j}{2}\right)\phi(2t - j - 2n)$$

$$= \sum_n a_{n,0}\sum_k \phi\left(\frac{k}{2} - n\right)\phi(2t - k)$$

$$= \sum_k f\left(\frac{k}{2}\right)\phi(2t - k).$$

The interchange of the two series is justified since the inner series is a convolution of two ℓ^1 sequences. This is all we need since the last series converges in $L^2(\mathbb{R})$.   □

We now turn to the Meyer type wavelets presented in Chapter 3. Their scaling functions ϕ were given by

$$\hat{\phi}(\omega) = \left\{ \int_{\omega-\pi}^{\omega+\pi} h(\zeta)d\zeta \right\}^{1/2}. \tag{9.18}$$

where h is a positive distribution whose smallest support interval is $[-\epsilon, \epsilon]$, $0 \leq \epsilon < \frac{\pi}{3}$.

LEMMA 9.5
Let $\hat{\phi}(\omega)$ satisfy (9.18); then it also satisfies (9.16).

PROOF   Since $\hat{\phi}$ is taken to be the positive square root in (9.18), we need only show that

$$|\hat{\phi}(\omega)|^2 = |\hat{\phi^*}(\omega)|^2 \left|\hat{\phi}\left(\frac{\omega}{2}\right)\right|^2. \tag{9.19}$$

Since $\hat{\phi}$ has support on $[-\pi - \epsilon, \pi + \epsilon]$, it follows that $\hat{\phi^*}$ has support on $\Lambda = \bigcup_k [-\epsilon + (4k-1)\pi, \epsilon + (4k+1)\pi]$. Thus on the support of $\hat{\phi}\left(\frac{\omega}{2}\right)$, $\hat{\phi} = \hat{\phi^*}$ and (9.19) becomes

$$|\hat{\phi}(\omega)|^2 = |\hat{\phi}(\omega)|^2 \left|\hat{\phi}\left(\frac{\omega}{2}\right)\right|^2.$$

Moreover, $\hat{\phi}\left(\frac{\omega}{2}\right) = 1$ on $[-2\pi + 2\epsilon, 2\pi - 2\epsilon] \supseteq [-\pi - \epsilon, \pi + \epsilon]$, the support of $\hat{\phi}$. Thus (9.19) holds.   □

We can also go in the opposite direction. We begin with (9.16) and try to get (9.18).

LEMMA 9.6
Let $\phi(t)$ be a scaling function satisfying the conditions of Lemma 9.3 and (9.16); let the support of $\hat{\phi}$ be a bounded interval; then there is a distribution $h, \hat{h}(0) = 1$, with support in an interval of length $\leq 2\pi/3$ contained in $[-\pi, \pi]$ such that

$$\int_{\omega-\pi}^{\omega+\pi} h \geq 0 \quad \text{and} \quad \hat{\phi}(\omega) = \left[\int_{\omega-\pi}^{\omega+\pi} h \right]^{1/2}.$$

PROOF   We first observe that the support of $\hat{\phi}$ must be a finite interval $[-a, b]$ where both a and b are positive. This follows from the fact that

$\hat{\phi}(0) = 1$. Since $\hat{\phi}$ is continuous its support contains a neighborhood of the origin.

The support of $\hat{\phi}_*^*$ is $\Lambda = \cup[-a + 4\pi k, b + 4\pi k]$ and hence if $b + a \geq 4\pi$, would be all of $\mathbb{R}$. But this is impossible since by (9.16) the support of $\hat{\phi}\left(\frac{\omega}{2}\right)$ would also have to be $[-a, b]$. We can say much more since by (9.16)

$$[-a, b] = [-2a, 2b] \cap \Lambda.$$

Hence $-a + 4\pi \geq 2b$ and $b - 4\pi \leq -2a$, which may be expressed as

$$a + 2b \leq 4\pi, \ 2a + b \leq 4\pi,$$

which may be added to obtain $a + b \leq 8\pi/3$.

On the other hand $2\pi \leq a + b$ since otherwise the orthogonality condition

$$\sum_k \left|\hat{\phi}(\omega + 2\pi k)\right|^2 = 1, \qquad \omega \in \mathbb{R}$$

would be violated.

Since on the support of $\hat{\phi}$, (9.16) becomes

$$\hat{\phi}(\omega) = \hat{\phi}(\omega)\hat{\phi}\left(\frac{\omega}{2}\right),$$

it follows that $\hat{\phi}\left(\frac{\omega}{2}\right) = 1$ on $[-a, b]$ or $\hat{\phi}(\omega) = 1$ on $\left[-\frac{a}{2}, \frac{b}{2}\right]$.
We now define $h(\omega)$ to be

$$h(\omega) := \begin{cases} -\dfrac{d}{d\omega}\left|\hat{\phi}(\omega + \pi)\right|^2, & 0 < \omega + \pi \\ 0, & \omega + \pi \leq 0 \end{cases} \qquad (9.20)$$

where the derivative is in general taken in the distribution sense.

It should be noticed that $h(\omega) = 0$ for $\omega < \pi - a$ or $\omega > b - \pi$. Furthermore, since $\left|\hat{\phi}(\omega + \pi)\right|^2 + \left|\hat{\phi}(\omega - \pi)\right|^2 = 1$ for $\pi - a < \omega < b - \pi$, it follows that

$$h(\omega) = \begin{cases} \dfrac{d}{d\omega}\left|\hat{\phi}(\omega - \pi)\right|^2, & \omega - \pi < 0 \\ 0, & \omega - \pi \geq 0. \end{cases}$$

From these two expressions we deduce that

$$\left|\hat{\phi}(\omega)\right|^2 = \int_{\omega - \pi}^{\omega + \pi} h(\zeta)\,dx \geq 0, \qquad \omega \in \mathbb{R},$$

which may be rewritten as the conclusion of the Lemma, since the length of the support of h is $b + a - 2\pi \leq 2\pi/3$. □

As a consequence of the hypothesis, $\left|\hat{\phi}(\omega)\right|^2$ must be continuous but not necessarily differentiable, so that h is not necessarily a function. To be consistent we must take $\hat{\phi}(\omega)$ to be the positive square root of $\left|\hat{\phi}(\omega)\right|^2$. Its inverse Fourier transform $\phi(t)$ is not necessarily real. However, $\hat{\phi}(\omega)$ is symmetric about $\frac{-a+b}{2}$ if $h(\omega)$ is symmetric about 0. Therefore, in this case $\phi(t)$ can be made real by shifting ω by an amount $\frac{-a+b}{2}$. Then $h(\omega)$ satisfies the condition needed except for the positivity. We add this as a hypothesis to get

THEOREM 9.7
Let $\phi(t)$ be a real, symmetric scaling function such that $\phi(t) = O\left(1 + |t|\right)^{-1-\epsilon}$, $\hat{\phi}(\omega) = O\left(1 + |\omega|\right)^{-1-\epsilon}$, $\epsilon > 0$, and $\hat{\phi}$ is nonincreasing for $\omega > 0$; then $\phi(t)$ is a Meyer type scaling function if and only if (i) $\phi(t)$ satisfies the double sampling property (9.15) and (ii) the support of $\hat{\phi}(\omega)$ is a bounded interval.

9.7 Cardinal scaling functions

In the previous examples, we considered wavelet subspaces in which the orthonormal basis is based on one function (a scaling function) and sampling on another (a cardinal function, i.e., a function f such that $f(n) = \delta_{on}$). In none of these examples are the two functions the same except in the case of Shannon wavelets where $\phi(t)$ satisfies both $\phi(n) = \delta_{on}$ and $\int_{-\infty}^{\infty} \phi(t)\phi(t-n)dt = \delta_{on}$.

However it is possible to construct new families of wavelets not previously considered that also satisfy both conditions, i.e., have a cardinal scaling function. One approach [X-Z] is based on the coefficients of the dilation equation

$$\phi(t) = \sum_k c_k \sqrt{2}\, \phi(2t - k). \tag{9.21}$$

The other [W-L] is based on the Fourier transform of $\phi(t)$, which satisfies

$$\hat{\phi}(\omega) = m_0\left(\frac{\omega}{2}\right)\hat{\phi}\left(\frac{\omega}{2}\right).$$

If $\phi(t)$ is to be a scaling function, then the coefficient $\{c_k\}$ must also satisfy (see Chapter 3)

$$\sum_k c_k c_{k-2n} = \delta_{on}, \qquad n \in \mathbb{Z} \tag{9.22a}$$

or equivalently

$$\left| m_0 \left(\frac{\omega}{2} \right) \right|^2 + \left| m_0 \left(\frac{\omega}{2} + \pi \right) \right|^2 = 1. \qquad (9.22b)$$

If it is to be a cardinal function then

$$\delta_{on} = \sum_k c_k \sqrt{2} \, \delta_{0,2n-k} = \sqrt{2} \, c_{2n}, \quad n \in \mathbb{Z}, \qquad (9.23a)$$

or equivalently,

$$m_0 \left(\frac{\omega}{2} \right) + m_0 \left(\frac{\omega}{2} + \pi \right) = 1. \qquad (9.23b)$$

These conditions are necessary but not sufficient for $\phi(t)$ to be a cardinal scaling function. Necessary and sufficient conditions may be given in terms of the Fourier transform of $\phi(t)$:

$$\sum_k \left| \hat{\phi}(\omega + 2\pi k) \right|^2 = 1, \qquad (9.22c)$$

$$\sum_k \hat{\phi}(\omega + 2\pi k) = 1. \qquad (9.23c)$$

That this last condition (9.23c) is equivalent to $\phi(n) = \delta_{on}$ may be shown by taking the Fourier series of the left side, which converges in the sense of $L^2(0, 2\pi)$. Its Fourier coefficients are then seen to be $\phi(-n) = \delta_{on}$, and the Fourier series is identically equal to 1.

Two other conditions needed are

$$\sum_k c_k = \sqrt{2}, \qquad \sum_k (-1)^k c_k = 0,$$

which correspond to

$$m_0(0) = 1, \qquad m_0(\pi) = 0.$$

The approach of Xia and Zhang uses the fact that the even coefficients are zero (by (9.23a)), except for c_0 and hence

$$m_0(\omega) = \frac{1}{2} + \frac{1}{\sqrt{2}} \sum_k c_{2k+1} e^{i\omega(2k+1)}$$

$$= \frac{1}{2} + \frac{e^{i\omega}}{2} H(2\omega)$$

where

$$H(\omega) = \sqrt{2} \sum_k c_{2k+1} e^{i\omega k} = \sum_k h_k e^{i\omega k}.$$

The conditions (9.22b) and (9.23b) are converted into conditions involving $H(\omega)$. They are

$$|H(\omega)|^2 = \left|\sum_k h_k e^{i\omega h}\right|^2 = \left|\sum_n \left(\sum_k h_{n-k} h_k\right) e^{i\omega n}\right|^2$$

$$= \left|\sum_n \delta_{on} e^{i\omega}\right|^2 = 1,$$

(9.22d)

and

$$H(0) = 1. \tag{9.23d}$$

Now it is just a matter of finding $H(\omega)$ satisfying these two conditions and then checking that these necessary conditions are also sufficient. The simplest example is

$$H_1(\omega) = e^{iM\omega} \frac{1 + a\,e^{i\omega}}{a + e^{i\omega}} \tag{9.24}$$

where M is an integer and $0 < |a| < 1$. The Fourier series of $H_1(\omega)$ is

$$H_1(\omega) = e^{i(M-1)\omega}(1 + a\,e^{i\omega})\sum_{n=0}^{\infty}(-a)^n e^{-in\omega}$$

$$= a\,e^{iM\omega} + \sum_{k=-\infty}^{M-1} (-a)^{M-k-1}(1 - a^2)e^{ik\omega}.$$

(9.25)

The coefficients can be converted back into c_k's and give us a Fourier series for $m_0(\omega)$ whose coefficients are exponentially decaying. A particular choice of a and M can be made to obtain an $m_0(\omega)$ corresponding to a moment condition $\int_{-\infty}^{\infty} t\psi(t)dt = 0$ (see Chapter 3). This is obtained by requiring $m'(\pi) = 0$, which leads to the condition

$$(2M + 3)a^2 + (4M + 2)a + 2M - 1 = 0$$

or

$$a = -\frac{2M - 1}{2M + 3}.$$

This lies between 0 and -1 for all $M \geq 1$ and is smallest when $M = 1$ and $a = -1/5$. This gives the most rapid convergence in (9.25).

THEOREM 9.8
Let $M = 1$ and $a = -1/5$ in $H_1(\omega)$ given by (9.24), let

$$m_0(\omega) = \frac{1}{\sqrt{2}}(1 + H_1(2\omega));$$

then $\phi(t)$ defined by

$$\hat{\phi}(\omega) = \prod_{k=1}^{\infty} m_0(2^{-k}\omega)$$

is a continuous cardinal scaling function.

PROOF To prove this we work backwards from conditions (9.22b) and (9.23b), which we know $m_0(\omega)$ satisfies, to (9.22c) and (9.23c) respectively. Part of the work has been done for us. Mallat [Ma], [D, p.182] showed that if $m_0(\omega)$ has no zeros for $|\omega| \leq \frac{\pi}{2}$, then $\phi(t)$ is an orthogonal scaling function. By direct calculation we find that

$$m_0(\omega) = \left(\frac{1+e^{i\omega}}{2}\right)^3 \cdot \frac{4(1-3e^{i\omega}+e^{2i\omega})}{1-5e^{2i\omega}}$$

and hence that the only zeros of $m_0(\omega)$ are at $(2k+1)\pi$, $k \in \mathbb{Z}$. Thus it remains only to be shown that $\phi(t)$ is continuous and that $\phi(n) = \delta_{on}$.

The continuity again follows from a general result [D, p.216]. In fact, by Lemma 7.1.1 given there, we have $\phi \in C^1$ because of the unexpected triple zero of $m_0(\omega)$ at $\omega = \pi$.

To prove that the cardinal condition holds, we imitate a proof of Cohen [D, p.183]. Let

$$g_k(\omega) := \prod_{j=1}^{k} m_0(2^{-j}\omega)\chi_{[-\pi,\pi]}(2^{-k}\omega), \quad k = 1,2,\cdots;$$

then $g_k(\omega) \to \hat{\phi}(\omega)$ dominatedly pointwise and

$$\mu_{n,k} := \frac{1}{2\pi}\int g_k(\omega)e^{i\omega n}d\omega \to \frac{1}{2\pi}\int \hat{\phi}(\omega)e^{i\omega n}d\omega = \phi(n).$$

But we have

$$\mu_{n,k} = \frac{1}{2\pi}\int_{-2^k\pi}^{2^k\pi} \prod_{j=1}^{k} m_0(2^{-j}\omega)e^{in\omega}d\omega$$

$$= \frac{2^k}{2\pi}\int_{-\pi}^{\pi} e^{in\xi 2^k} \prod_{j=1}^{k-1} m_0(2^j\xi)m_0(\xi)d\xi$$

$$= \frac{2^k}{2\pi}\int_{0}^{\pi} e^{in\xi 2^k} \prod_{j=1}^{k-1} m_0(2^j\xi)(m_0(\xi)+m_0(\xi-\pi))d\xi$$

since $e^{in\xi 2^k} \prod_{j=1}^{k-1} m_0(2^j\xi)$ is periodic π. Hence

$$\mu_{n,k} = \frac{2^k}{2\pi} \int_0^\pi e^{in\xi 2^k} \prod_{j=1}^{k-1} m_0(2^j\xi)\,d\xi$$

$$= \frac{2^{k-1}}{2\pi} \int_0^{2\pi} e^{in\xi 2^{k-1}} \prod_{j=0}^{k-2} m_0(2^j\xi)\,d\xi$$

$$= \mu_{n,k-1}.$$

Since

$$\mu_{n,1} = \frac{1}{2\pi} \int_{-2\pi}^{2\pi} m_0(2^{-1}\omega)e^{in\omega}\,d\omega$$

$$= \frac{1}{2\pi} \int_0^{2\pi} \left(m_0\left(\frac{k}{2}\right)e^{in\omega} + m_0\left(\frac{\omega}{2} - \pi\right)e^{in(\omega-2\pi)} \right)\,d\omega$$

$$= \frac{1}{2\pi} \int_0^{2\pi} e^{in\omega}\,d\omega = \delta_{on},$$

it follows that

$$\delta_{on} = \lim_{k\to\infty} \mu_{n,k} = \phi(n). \qquad \square$$

Other examples of cardinal scaling functions using a similar construction may be found in [X-Z]. However, these are not bandlimited as are the Shannon scaling functions. It is shown in [W-L] that a family of cardinal scaling functions that are bandlimited and that are rapidly decreasing can also be constructed. They are based on the following lemma [W-L], which does not require that $\phi \in S_r$.

LEMMA 9.9
Let $\phi(x) \in L^2(\mathbb{R})$. Then $\phi(x)$ is an orthonormal cardinal scaling function if

(1) $\sum_k \left|\hat{\phi}(\omega + 2\pi k)\right|^2 = 1$ *a.e.*,

(2) there exists a 2π-periodic and Lipschitz continuous function $m_0(\omega) \in L^2(-\pi, \pi)$ such that $\hat{\phi}(\omega) = m_0\left(\frac{\omega}{2}\right) \cdot \hat{\phi}\left(\frac{\omega}{2}\right)$ *a.e.*,

(3) $\hat{\phi}(\omega)$ is continuous at 0 and $\hat{\phi}(0) = 1$,

(4) $\sum_k \hat{\phi}(\omega + 2\pi k) = 1$ *a.e.*

To construct our scaling function, we shall prove a necessary condition first. The Theorem 9.11 then gives the desired scaling functions and the mother wavelets.

LEMMA 9.10

If $\phi(t) \in (L^2 \cap L^1)(I\!R)$ is an orthonormal cardinal scaling function with supp $\hat{\phi} \subseteq [-\pi - \epsilon, \pi + \epsilon]$ $(0 < \epsilon \leq \frac{\pi}{3})$ and $\hat{\phi}(\omega) = \overline{\hat{\phi}(-\omega)}$, then $\hat{\phi}(\omega)$ must have the form

$$\hat{\phi}(\omega) = \frac{1}{2} + \frac{1}{2}e^{i\theta(\omega)}, \quad |\omega| < 2\pi$$

for some real function $\theta(\omega)$. Furthermore $\theta(\omega)$ satisfies
(a)' $\theta(-\omega) = -\theta(\omega) + 2\pi k,$
(b)' $|\theta(\omega)| = (2k + 1)\pi$ *for* $|\omega| \geq \pi + \epsilon,$
(c)' $\theta(\omega) + \theta(2\pi - \omega) = (2k + 1)\pi.$

PROOF   Since $\phi(n) = \delta_{no}$, it follows that (9.23c) holds and the orthonormality of $\{\phi(t - n)\}$ implies that (9.22c) holds.

The condition supp $\hat{\phi} \subseteq [-\pi - \epsilon, \pi + \epsilon]$ ensures that the series in these equations reduces to two terms for each $\omega \in R$. In particular, for $0 \leq \omega < 2\pi$, one has

$$\hat{\phi}(\omega) + \hat{\phi}(\omega - 2\pi) = 1 \quad a.e. \tag{9.26}$$

and

$$|\hat{\phi}(\omega)|^2 + |\hat{\phi}(\omega - 2\pi)|^2 = 1 \quad a.e. \tag{9.27}$$

The general solution to (9.26) and (9.27) must have the form

$$\hat{\phi}(\omega) = \frac{1}{2} + \frac{1}{2}e^{i\theta(\omega)} \tag{9.28}$$

for some real function $\theta(\omega)$. Similarly (9.28) holds for $-2\pi < \omega \leq 0$.

By using (9.28), (a)' follows from $\hat{\phi}(\omega) = \overline{\hat{\phi}(-\omega)}$, (b)' holds since supp $\hat{\phi} \subseteq [-\pi - \epsilon, \pi + \epsilon]$, and (c)' holds because of (9.26) and (a)'.   □

Next we shall choose $\theta(\omega)$ such that $\hat{\phi}(\omega) = \frac{1}{2} + \frac{1}{2}e^{i\theta(\omega)}$ satisfies all conditions of Lemma 9.10.   Let $F = \{\alpha \in L^1(R) | \alpha(\omega) \geq 0,$ supp $\alpha \in [-\epsilon, \epsilon], \alpha(-\omega) = \alpha(\omega)$ and $\int_{-\epsilon}^{\epsilon} \alpha(\omega)d\omega = \pi\}$ for some $0 < \epsilon \leq \frac{\pi}{3}$. Define, for such an $\alpha \in F$,

$$\theta(\omega) := \int_{-\omega-\pi}^{\omega-\pi} \alpha(\xi)d\xi \tag{9.29}$$

and

$$\hat{\phi}(\omega) := \frac{1}{2} + \frac{1}{2}e^{i\theta(\omega)}.$$

Then the corresponding $\phi(t)$ is the desired function for any $\alpha(\omega) \in F$. In fact we have

THEOREM 9.11
For each $\alpha(\omega) \in F$, the function $\phi(t)$ defined by (9.18) and (9.19) is a real bandlimited orthonormal cardinal scaling function, and the corresponding mother wavelet $\psi(t)$ is given by

$$\psi(t) = 2\phi(2t - 1) - \phi\left(\frac{1}{2} - t\right).$$

PROOF Since $\alpha(\omega) \in F$, it is easy to see that

$$\theta(\omega) = \int_{-\omega-\pi}^{\omega-\pi} \alpha(\xi)d\xi$$

satisfies

    (i) $\theta(-\omega) = -\theta(\omega)$,
    (ii) $|\theta(\omega)| = \pi$ for $|\omega| \geq \pi + \epsilon$,
    (iii) $\theta(\omega) = 0$ for $|\omega| \leq \pi - \epsilon$,
    (iv) $\theta(\omega) + \theta(2\pi - \omega) = \pi$ for $0 \leq 2\pi$.

Furthermore, it can be shown that the function $\hat{\phi}(\omega) = \frac{1}{2} + \frac{1}{2}e^{i\theta(\omega)}$ has the following properties:

    (i)' $\hat{\phi}(\omega) = \hat{\phi}(-\omega)$,
    (ii)' $\hat{\phi}(\omega) = 0$ for $|\omega| \geq \pi + \epsilon$,
    (iii)' $\hat{\phi}(\omega) = 1$ for $|\omega| \leq \pi - \epsilon$,
    (iv)' $\hat{\phi}(\omega) + \hat{\phi}(\omega + 2\pi) + \hat{\phi}(\omega - 2\pi) = 1$ for $|\omega| < 2\pi$,
    (v)' $|\hat{\phi}(\omega)|^2 + |\hat{\phi}(\omega + 2\pi)|^2 + |\hat{\phi}(\omega - 2\pi)|^2 = 1$ for $|\omega| < 2\pi$,
    (vi)' $\hat{\phi}(\omega) = \sum_k \hat{\phi}(\omega + 4\pi k) \cdot \hat{\phi}\left(\frac{\omega}{2}\right)$ for $\omega \in R$.

In fact, the conclusion (i)'–(iii)' follow from (i)–(iii) directly; (iv)' and (v)' do from (iv), by considering two cases $0 \leq \omega < 2\pi$ and $-2\pi < \omega \leq 0$; for the proof of (iv)', the real line $\mathbb{R}$ may be separated into three disjoint sets: $|\omega| \geq \pi + \epsilon$, $\pi - \epsilon \leq |\omega| \leq \pi + \epsilon$ and $|\omega| \leq \pi - \epsilon$.

Define $m_0\left(\frac{\omega}{2}\right) := \sum_k \hat{\phi}(\omega + 4\pi k)$. Then all conditions except (4) in Lemma 9.9 are satisfied because of (ii)', (iii)', (v)' and (vi)'.

As in the beginning of the proof of Lemma 9.10, it can be shown that $\phi(t)$ is a cardinal function by (iv)'. Obviously (i) implies that $\phi(t)$ is real.

Next one needs to calculate $\psi(t)$ by the general formula

$$\hat{\psi}(\omega) = e^{-i\frac{\omega}{2}} \overline{m_0\left(\frac{\omega}{2} + \pi\right)} \hat{\phi}\left(\frac{\omega}{2}\right).$$

Since $m_0\left(\frac{\omega}{2}\right) = \sum_k \hat{\phi}(\omega + 4\pi k)$ with supp $\hat{\phi} \in \left[-\frac{4\pi}{3}, \frac{4\pi}{3}\right]$ $\left(0 < \epsilon \leq \frac{\pi}{3}\right)$,

$\overline{m_0\left(\frac{\omega}{2} + \pi\right)} \hat{\phi}\left(\frac{\omega}{2}\right)$ reduces to $\overline{\hat{\phi}(\omega + 2\pi) + \hat{\phi}(\omega - 2\pi)} \cdot \hat{\phi}\left(\frac{\omega}{2}\right)$. Therefore by $\hat{\phi}(\omega) = \frac{1}{2} + \frac{1}{2}e^{i\theta(\omega)}$, one has

$$\hat{\psi}(\omega) = e^{-i\frac{\omega}{2}} \left[1 + \frac{e^{-i\theta(\omega+2\pi)} + e^{-i\theta(\omega-2\pi)}}{2} \right] \cdot \frac{1 + e^{i\theta(\frac{\omega}{2})}}{2}.$$

Furthermore it can be shown that

$$\hat{\psi}(\omega) = e^{-i\frac{\omega}{2}} \frac{1 + e^{i[\pi-\theta(\omega)]\chi(\omega)}}{2} \cdot \frac{1 + e^{i\theta(\omega)}}{2}$$

by using the properties (i)–(iv) of $\theta(\omega)$, where $\chi(\omega)$ is the characteristic function on $[-2\pi, 2\pi]$.

To complete the proof, one needs to show

$$\psi(t) = 2\phi(2t-1) - \phi\left(\frac{1}{2} - t\right).$$

Since $\hat{\phi}(\omega) = m_0\left(\frac{\omega}{2}\right)\hat{\phi}\left(\frac{\omega}{2}\right)$ and $m_0\left(\frac{\omega}{2}\right)$ vanishes for those ω whose $\hat{\phi}\left(\frac{\omega}{2}\right)$ is not real (which is possible only on the interval $[2\pi - 2\epsilon, 2\pi + 2\epsilon]$), one has that

$$\hat{\phi}(\omega) = m_0\left(\frac{\omega}{2}\right) \cdot \overline{\hat{\phi}\left(\frac{\omega}{2}\right)}.$$

By (9.23b) one has $m_0(\omega) + m_0(\omega + \pi) = 1$ and hence

$$\hat{\psi}(\omega) = e^{-i\frac{\omega}{2}} \overline{m_0\left(\frac{w}{2} + \pi\right)} \hat{\phi}\left(\frac{w}{2}\right) = e^{i\frac{\omega}{2}} \overline{\left[1 - m_0\left(\frac{\omega}{2}\right)\right]} \hat{\phi}\left(\frac{\omega}{2}\right)$$

$$= e^{-i\frac{\omega}{2}} \left[\hat{\phi}\left(\frac{\omega}{2}\right) - \overline{m_0\left(\frac{\omega}{2}\right)} \hat{\phi}\left(\frac{\omega}{2}\right)\right] = e^{-i\frac{\omega}{2}} \left[\hat{\phi}\left(\frac{\omega}{2}\right) - \overline{\hat{\phi}(\omega)}\right].$$

It follows that $\psi(t) = 2\phi(2t-1) - \phi\left(\frac{1}{2} - t\right)$, by taking the inverse Fourier transform.   □

9.8   Problems

1. Find the Riesz basis $\{q(x,n)\}$ of V_0 for the Daubechies wavelets.

2. For the Franklin wavelets, find a biorthogonal sequence $\{g(t-n)\}$ in V_0 to the Riesz basis $\{S(t-n)\}$ where $S(t)$ is the hat function. (Hint: Use the same trick as in orthogonalization and the Fourier transform $\sum_k \hat{S}(w + i\pi k)\hat{g}(w + 2\pi k) = 1$.)

3. For the Franklin wavelets, shifted sampling is impossible for $a = \frac{1}{2}$. Show that the condition

$$\sum \phi(a+n)e^{-iwn} \neq 0$$

fails in this case. (Use the original B-spline $\theta_2(x)$ instead of ϕ.)

4. Show that for the B-spline $\theta_2(x)$ ($=$ hat function), $\theta_2(a)+\theta_2(a+1) = 1$ for $0 < a < 1$. Use this to show that the criterion holds for such $a \neq \frac{1}{2}$ and hence there is a shifted sampling theorem.

5. Let $h(\xi) = 2\epsilon$, $|\xi| < \epsilon \le \frac{\pi}{3}$ and $h(\xi) = 0$, $|\xi| \ge \epsilon$; let $\hat{\phi}(w)$, given by (9.18), be the associated Meyer wavelets. Indicate that (9.16) holds in this case by sketching the graphs of the functions involved.

6. Find an upper bound on the aliasing error for $f \in V_2$ when the sampling series is in V_0.

7. Expand $f(t) = \frac{\sin \pi t/2}{t}$ in terms of the Shannon sampling theorem. Is $f \in V_0$? V_1?

10

Translation and Dilation Invariance in Orthogonal Systems

One of the reasons for taking orthogonal expansions of functions is that operations on the functions may be converted to simpler operations on the coefficients. Thus, in the example of the Sturm-Liouville systems, the differential operator is converted into multiplication by the eigenvalues.

Many of the operations that are useful in analysis are based on translation and dilation operators, the same operators that appear repeatedly in wavelet theory. In this chapter we study these two operators and their effect on the partial sums of the expansions. We first consider trigonometric systems and orthogonal polynomials and then a particular wavelet system in which everything works well. However, in most wavelet systems translation invariance fails, and we therefore consider a weaker concept that works more often. In fact, the wavelets that satisfy this "weak" translation property are characterized as a type of Meyer wavelet. The same ones also satisfy a weak dilation invariance property.

10.1 Trigonometric system

An important property of the trigonometric system of orthogonal functions is its invariance under arbitrary translations. Indeed if $f(t)$ is periodic 2π with Fourier series

$$f(t) \sim \sum_n c_n e^{int}, \tag{10.1}$$

then $f(t - \alpha)$ has Fourier series

$$f(t - \alpha) \sim \sum_n (c_n e^{-in\alpha}) e^{int}. \tag{10.2}$$

In particular, the number of terms in the partial sums of the first functions is invariant under translation, the moduli of the coefficients are the same $|c_n| = |c_n e^{-in\alpha}|$, and the translations in the space V_m spanned by $1, e^{\pm it}, \cdots, e^{\pm imt}$, are again in this space.

Things are not so nice when it comes to dilation since for arbitrary $\beta > 0$, $f(\beta t)$ is not even periodic. The only regularity obtained is when $\beta = p$, a positive integer. Then the partial sum $f_m(t)$ in V_m will be mapped into V_{pm}.

Other common operators arise from translations. For example, differentiation of $f(t)$ leads to

$$\sum_n in\, c_n\, e^{int},$$

and V_m maps into itself under this operation. The same is true for convolution. If g is periodic 2π and in $L^1(0, 2\pi)$, then the Fourier series of the convolution is

$$f * g(t) = \int_0^{2\pi} f(t-s)g(s)ds = 2\pi \sum_n c_n d_n e^{int},$$

and hence for $f_m \in V_m$,

$$f_m * g \in V_m.$$

10.2   Orthogonal polynomials

When we consider orthogonal polynomials, things are not quite so simple. If $f(t)$ has an orthogonal polynomial expansion

$$f(t) \sim \sum_{n=0}^{\infty} c_n p_n(t), \tag{10.3}$$

then

$$f(t - \alpha) \sim \sum_n c_n \sum_{k=0}^{n} a_{nk}^{\alpha} p_k(t)$$
$$= \sum_k \left(\sum_{n=k}^{\infty} c_n a_{nk}^{\alpha} \right) p_k(t). \tag{10.4}$$

where a_{nk}^{α} are the expansion coefficients of $p_n(t - \alpha)$ in terms of p_k. If again we denote by V_m the space spanned by $p_0, p_1, \cdots, p_m$, then V_m maps into itself under translation. However, in the case of the entire

series as opposed to the partial sums, there is no guarantee that the
series of $f(t - \alpha)$ converges (or indeed that $f(t - \alpha)$ makes sense over
the interval of orthogonality).

The same is true for arbitrary positive dilations. $p_n(\beta t)$ is again a
polynomial of degree n and hence has an expansion in terms of p_0, p_1,
$\cdots$, p_n. Thus, again V_m maps into itself under dilations. The differential
operator maps V_m into V_{m-1} while the convolution operator maps V_m
into V_m provided $g * p$ exists.

10.3   An example where everything works

The wavelet prototype II discussed in Chapter 1 has certain desirable
properties with respect to dilation and translation. In particular, it has
the translation invariance properties of Fourier series. If V_m is a subspace
in the multiresolution analysis (MRA), then $T_\alpha f \in V_m$ whenever $f \in
V_m$. This follows from the fact that for $f \in V_o$, say,

$$\hat{f}(\omega) = F(\omega)\hat{\phi}(\omega) \tag{10.5}$$

where $F(\omega)$ is 2π periodic and $\hat{\phi}(\omega) = \chi_{[-\pi,\pi]}(\omega)$. The Fourier transform
of $f(t - \alpha)$ is $e^{-i\omega\alpha}F(\omega)\hat{\phi}(\omega)$, which by taking its periodic extension and
multiplying by $\hat{\phi}(\omega)$ can be put in the same form as (10.5).

The dilation $D_\beta f$ (defined by $(D_\beta f(x) = f(\beta x))$ transforms into
$\hat{f}\left(\frac{\omega}{\beta}\right)\frac{1}{\beta}$. This will belong to $\hat{V}_o$ provided $0 < \beta \leq 1$, since then its
support will be contained in $[-\pi, \pi]$. For dilations with $0 < \beta \leq 2$,
$D_\beta f \in V_1$, and by induction $D_\beta f \in V_m$ whenever $0 < \beta \leq m$. Thus all
positive dilations of $f \in V_m$ belong to one of the spaces in the MRA.

As a consequence of the translation property, all derivatives of $f \in V_m$
are in V_m and hence any differential operator with constant coefficients
maps V_m into itself. This can be extended to operators with polynomial
coefficients provided that $f \in V_m$ is sufficiently rapidly decreasing.

The convolution of $f \in V_m$ with any $L^1(\mathbb{R})$ function g is again in V_m
since the Fourier transform $\hat{g}(\omega)$ is continuous, and the Fourier transform
of the convolution is

$$\hat{g}(\omega)\hat{f}(\omega),$$

which has support in $[-\pi, \pi]$ and, since $\hat{g}$ is bounded, belongs to
$L^2[-\pi, \pi]$. Thus, $\hat{g}(\omega)\hat{f}(\omega)$ may be put in the form (10.5). In this example, therefore, considerable analysis can be done without ever leaving
the space V_m. Unfortunately, this is not typical of wavelets.

10.4   An example where nothing works

The wavelet prototype I of Chapter 1 is at the opposite extreme. In this case $\phi(t)$ is the characteristic function of the unit interval. Clearly $\phi(t - \alpha)$ where α is an irrational number does not belong to V_o. In fact, it belongs to <u>no</u> V_m in the associated MRA since $f \in V_m$ is piecewise constant with possible jumps at $2^{-m}n$, $n \in \mathbb{Z}$.

For dilations we have a similar negative result. The dilation $D_\beta\phi$ belongs to none of the V_m when β is an irrational number.

We might expect the convolution of ϕ with some function g to be in V_o in spite of the failure of translation.

But $g * \phi$ is given by

$$\int_{-\infty}^{\infty} g(t)\phi(x - t)dt = \int_{x-1}^{x} g(t)dt,$$

which is continuous and if it is to be in V_o, a constant. Then the derivative must be $0 = g(x) - g(x - 1)$. This makes g periodic and hence not in $L^1(\mathbb{R})$ unless it is zero.

One might ask which of these two examples is more typical. Unfortunately, it is the latter, since as shown by Madych [Md], the only scaling functions for which V_o is translation invariant are those whose Fourier transform is a characteristic function of some set. The only standard such one is our prototype II.

10.5   Weak translation invariance

In this section we modify the definition of translation invariance. Rather than requiring that each subspace in a multiresolution analysis $\{V_m\}$ be invariant we shall require only that translations be in the next space up the ladder V_{m+1}. This is almost as good and includes many additional wavelet bases, in particular the Meyer wavelets [W8].

We first study weakly translation invariant MRA in general and find necessary and sufficient conditions for them. With an added connectivity condition, these are exactly the MRA of Meyer type wavelets. These are shown to be invariant as well under other operators including dilation, differentiation, and convolution in the next section.

DEFINITION 10.1
Let $\{V_m\}$ be a multiresolution analysis of $L^2(\mathbb{R})$. Then $\{V_m\}$ is said to be <u>weakly translation invariant</u> if the translation operator, T_α, $\alpha \in \mathbb{R}$, maps V_m into V_{m+1}.

Any translation invariant MRA will be weakly translation invariant since $V_m \subseteq V_{m+1}$. This includes the Shannon wavelets $(\phi(t) = \frac{\sin \pi t}{\pi t})$ but no other nonpathological cases [Md]. All of the translation invariant MRA's have $\phi \notin L^1(\mathbb{R})$ and thus give poor time localization.

The prototype of the weakly translation invariant MRA is the family of Meyer wavelets [L-M] that have $\phi(t)$ rapidly decreasing. These are quite useful in some applications, such as $1/f$ processes, in which the frequency is known only over a finite range.

PROPOSITION 10.1
Let $\{V_m\}$ be a weakly translation invariant MRA of $L^2(\mathbb{R})$ with scaling function ϕ; let $\Omega = supp \, \hat{\phi}$.
Then

(a) $\Omega \subset 2\Omega$
(b) $\Omega \bigcap \{\Omega + 4\pi k\} \simeq \Phi, \; k \in \mathbb{Z}, \; k \neq 0$
(c) $\bigcup\limits_{k \in \mathbb{Z}} \{\Omega + 2\pi k\} \simeq \mathbb{R}$
(d) $\bigcup\limits_{j \in \mathbb{Z}} (2^j \Omega) \simeq \mathbb{R}$
(e) *There is a distribution (or function) h with support on $\{\Omega + \pi\} \bigcap \{\Omega - \pi\}$ such that*

$$\frac{d}{dw} \left| \hat{\phi}(w) \right|^2 = \begin{cases} h(w + \pi), & w \in \Omega - \{\Omega + 2\pi\} \\ -h(w - \pi), & w \in \Omega - \{\Omega - 2\pi\}, \end{cases}$$

(f) $\left| \hat{\phi}(w) \right| = 1, \; w \in \Omega - \{\Omega + 2\pi\} - \{\Omega - 2\pi\}.$

PROOF Since $\hat{\phi}(w) = m_0 \left(\frac{w}{2}\right) \hat{\phi} \left(\frac{w}{2}\right)$, where m_0 is a 2π-periodic function and supp $\hat{\phi} \left(\frac{w}{2}\right) = 2\Omega$, (a) clearly follows.

For (b), we observe that $\widehat{T_\alpha \phi}(w) = e^{-iw\alpha} \hat{\phi}(w)$, and since $T_\alpha \phi \in V_1$ by the weak translation invariance,

$$\widehat{T_\alpha \phi}(w) = m_\alpha \left(\frac{w}{2}\right) \hat{\phi} \left(\frac{w}{2}\right),$$

where m_α is also 2π periodic. Hence we have

$$\left(m_\alpha \left(\frac{w}{2}\right) - e^{-iw\alpha} m_0 \left(\frac{w}{2}\right) \right) \hat{\phi} \left(\frac{w}{2}\right) = 0$$

and by shifting w by $w\pi k$ we find

$$m_\alpha \left(\frac{w}{2}\right) - e^{-(w + 4\pi k)\alpha} m_0 \left(\frac{w}{2}\right) = 0, \quad w \in \{2\Omega + 4\pi k\},$$

since the support of $\hat{\phi} \left(\frac{w + 4\pi k}{2}\right)$ is $\{2\Omega + 4\pi k\}$.

Since $\{\Omega + 4\pi k\} \subset \{2\Omega + 4\pi k\}$ by (a) we have that

$$\left(e^{-i(w+4\pi k)\alpha} - e^{-iw\alpha}\right) m_0\left(\frac{w}{2}\right) = 0, \quad w \in \Omega \bigcap \{\Omega + 4\pi k\}.$$

Moreover, $m_0\left(\frac{w}{2}\right) \neq 0$ for $w \in \Omega$ and by the periodicity for $w \in \{\Omega + 4\pi k\}$. Hence $e^{-i4\pi k\alpha} = 1$ or $\Omega \bigcap \{\Omega + 4\pi k\} = \phi$ for $k \neq 0$.

The proof of (c) follows immediately from the orthogonality condition

$$\sum_k \left|\hat{\phi}(w + 2\pi k)\right|^2 = 1. \tag{10.6}$$

If (d) does not hold, then there is set A of finite positive measure such that $A \subseteq (\mathbb{R}) - \bigcup_j 2^j \Omega$. Let $\hat{f}(w) = \chi_A(w)$; then $f \in L^2(\mathbb{R})$, $\hat{f} = 0$ on $2^m\Omega$ for each $m \in \mathbb{Z}$, and $f_m = P_m f = 0$ where P_m is the projector on V_m. This contradicts the fact that $\bigcup V_m$ is dense in $L^2(\mathbb{R})$.

For conditions (e) and (f) we use the fact that (10.6) contains at most two nonzero terms for each $w \in \mathbb{R}$, which follows from (b). We define $h(w)$ as the distribution $\in S'$ given by

$$h(w) = \begin{cases} \dfrac{d}{dw}\left|\hat{\phi}^2(w - \pi)\right|, & w \in \{\Omega + \pi\} \setminus \{\Omega + 3\pi\} \\ 0, & \text{otherwise} \end{cases} \tag{10.7}$$

and use the fact that

$$\left|\hat{\phi}^2(w + \pi)\right| + \left|\hat{\phi}^2(w - \pi)\right| = 1$$

for $w \in (\{\Omega + \pi\} \bigcup \{\Omega - \pi\}) \setminus (\{\Omega + 3\pi\} \bigcup \{\Omega - 3\pi\})$.

Hence

$$\frac{d}{dw}\left|\hat{\phi}(w + \pi)\right|^2 = -h(w), \quad w \in \{\Omega - \pi\} \setminus \{\Omega - 3\pi\}. \tag{10.8}$$

The support of $h(w)$ is contained in $\{\Omega + \pi\}$ by (10.7) and in $\{\Omega - \pi\}$ by (10.8). Their intersection is not empty by (10.6). The remaining conclusion in (e) follows by shifting (10.7) and (10.8).

The conclusion (f) also follows easily from (10.6). $\square$

This result should be compared to those in [Md], section 3 in which translation invariance is considered. In particular, it is shown there that $|\hat{\phi}|$ is the characteristic function of Ω and hence is not continuous. Therefore, $\phi \notin L^1(\mathbb{R})$. In our case $\hat{\phi}$ can be taken to be C^∞ with compact support and hence ϕ will be rapidly decreasing.

When Ω is a connected set, then the results can be made more exact. We assume also that Ω is symmetric about the origin. It can be made so by a simple shift of w so that nonsymmetric results are easily obtained from the symmetric ones.

COROLLARY 10.2
Let $\{V_m\}$ be a MRA as in Proposition 10.1 and suppose $\Omega = supp\ \hat{\phi}$ is a symmetric connected set. Then there is an $0 \le \epsilon \le \frac{\pi}{3}$ and a distribution h, with support in $[-\epsilon, \epsilon]$ such that

(i) $|\hat{\phi}(w)|^2 = \int\limits_{w-\pi}^{w+\pi} h,$

(ii) $supp\ \hat{\phi} \subseteq [-\pi - \epsilon, \pi + \epsilon],$

(iii) $|\hat{\phi}(w)| = 1, \quad |w| \le \pi - \epsilon.$

PROOF   Since Ω is connected it must be an interval $[a, b]$ which by (d) must contain a neighborhood of 0. It must also be bounded by (b) and in fact $b \le a + 4\pi$, or

$$b - a \le 4\pi. \tag{10.9}$$

Similarly by (c) it must be bounded below by

$$2\pi \le b - a. \tag{10.10}$$

The intersection of $\{\Omega + \pi\}$ and $\{\Omega - \pi\}$ is $[a + \pi, b - \pi]$, which cannot be empty by (10.10), and the set on which $|\hat{\phi}(w)| = 1$ is $[b - 2\pi, a + 2\pi]$ by (f).

Since Ω is symmetric $a = -b$ and ϵ may be taken as $\epsilon = b - \pi$. Then conclusions (ii) and (iii) follow. Clearly $0 \le \epsilon \le \pi$ by (10.9) and (10.10). We must show that $\epsilon \le \frac{\pi}{3}$.

To do so, we return to the dilation equation. The support of $\hat{\phi}\left(\frac{w}{2}\right)$ is $[-2\pi - 2\epsilon, 2\pi + 2\epsilon]$, hence $m_0\left(\frac{w}{2}\right)$ must be zero for $\pi + \epsilon < |w| \le 2\pi + 2\epsilon$. Since it is periodic 4π as a function of w, it follows that in every interval of length 4π, $m_0\left(\frac{w}{2}\right) = 0$ on a set of measure $\ge \pi + \epsilon$. In particular, on the interval $[-2\pi, 2\pi]$ the support of $m_0\left(\frac{w}{2}\right)$ (or of $\hat{\phi}(w)$) is of length $\le 3\pi - \epsilon$. Since Ω has length $2\pi + 2\epsilon$, it follows that $2\pi + 2\epsilon \le 3\pi - \epsilon$ or $\epsilon \le \frac{\pi}{3}$. The conclusion (i) then follows from (e). $\square$

It is clear that h need not necessarily be positive. Take for example

$$|\hat{\phi}(w)|^2 = \begin{cases} 1 & |w| \le \pi - \epsilon \\ 1/4 & \pi - \epsilon \le |w| \le \pi \\ 3/4 & \pi \le |w| \le \pi + \epsilon \\ 0 & \text{o.w.} \end{cases}$$

then $h(w) = \frac{3}{4}\delta(w + \epsilon) - \frac{1}{2}\delta(w) + \frac{3}{4}\delta(w - \epsilon)$. This may also be made into a smooth version by approximating the δ's by a smooth function.

However, since $|\hat{\phi}(w)|^2 \geq 0$, the mean of h over any interval of length 2π must be nonnegative. With this added condition the conclusions of the Corollary are sufficient as well as necessary.

PROPOSITION 10.3
Let the function $\phi(t)$ be given by

$$\hat{\phi}(w) = \left\{ \int_{w-\pi}^{w+\pi} h \right\}^{1/2}$$

where h is a distribution with support in $\left[-\frac{\pi}{3}, \frac{\pi}{3}\right]$ such that $h^*\chi_{[-\pi,\pi]}$ is a nonnegative bounded measurable function, and $\int_{-\infty}^{\infty} h = 1$. Then $\phi(t)$ is a scaling function and its associated MRA $\{V_m\}$ is weakly translation invariant.

PROOF We first consider some of the properties of $|\hat{\phi}(w)|^2$. It's clear that

(i) $\text{supp}|\hat{\phi}(w)|^2 \subseteq \left[-\frac{4\pi}{3}, \frac{4\pi}{3}\right]$,

(ii) $|\hat{\phi}(w)| = 1$ for $|w| \leq \frac{2\pi}{3}$,

(iii) $\sum_k |\hat{\phi}(w + 2\pi k)|^2 = \int_{-\infty}^{\infty} h = 1, \ a.e..$

In order to show that ϕ is a scaling function we must show that the dilation equation holds. If we define

$$m_0 \left(\frac{w}{2}\right) := \sum_k \hat{\phi}(w + 4\pi k)$$

then $m_0\left(\frac{w}{2}\right) = 0$ for $\frac{4\pi}{3} < |w| < \frac{8\pi}{3}$ and hence

$$\hat{\phi}(w) = m_0\left(\frac{w}{2}\right) \hat{\phi}\left(\frac{w}{2}\right)$$

holds. Since by condition (iii), ϕ is orthogonal to its translates as well, it is a scaling function.

Now let $f \in V_0$ and $f_\alpha = T_\alpha f$ be its translate by an amount α. Then

$$\hat{f}_\alpha(w) = \sum_n a_n e^{-iwn} e^{-iw\alpha} \hat{\phi}(w)$$

$$= F(w) e^{-iw\alpha} \hat{\phi}(w)$$

$$= F(w) e^{-iw\alpha} \theta(w) \hat{\phi}(w)$$

where $\theta(w) \in C^\infty$, supp $\hat{\phi}(w) \subseteq$ supp $\theta(w) \subseteq (-2\pi, 2\pi)$, and $\theta(w) = 1$ for $w \in$ supp $\hat{\phi}(w)$. We form the 4π periodic function

$$G_\alpha(w) := \sum_{k=-\infty}^{\infty} e^{-i(w+4\pi k)\alpha}\theta(w + 4\pi k).$$

Then $\hat{f}_\alpha$ becomes

$$\hat{f}_\alpha(w) = F(w)G_\alpha(w)\hat{\phi}(w) = F(w)G_\alpha(w)m_0\left(\frac{w}{2}\right)\hat{\phi}\left(\frac{w}{2}\right)$$

where $FG_\alpha m_0\left(\frac{\cdot}{2}\right) \in L^2(-2\pi, 2\pi)$ and is periodic 4π. Hence if its Fourier series is

$$F(w)G_\alpha(w)m_0\left(\frac{w}{2}\right) \sim \sum_{n=\infty}^{\infty} b_n e^{-iwn/2},$$

then

$$f_\alpha(t) = \sum_{n=\infty}^{\infty} b_n 2\phi(2t - n), \quad \{b_n\} \in \ell^2.$$

Thus $f_\alpha \in V_1$, and $\{V_m\}$ is a weakly translation invariant MRA.   □

These MRA's have many other nice properties (see below). Some of the examples have been discussed in Chapter 3 where they were generated by Meyer type wavelets (Example 7), which are slightly less general in that $h \geq 0$.

Lemarie and Meyer defined a wavelet whose scaling function [D, p.137] turned out to be

$$\hat{\phi}(w) = \begin{cases} 1, & |w| \leq \frac{2\pi}{3} \\ \cos\left[\frac{\pi}{2}\nu\left(\frac{3}{2\pi}|w| - 1\right)\right], & \frac{2\pi}{3} \leq |w| \leq \frac{4\pi}{3} \\ 0, & \text{otherwise} \end{cases}$$

where $\nu(x)$ is a C^k function satisfying

$$\nu(x) = \begin{cases} 0, & x \leq 0 \\ 1, & 1 \leq x \end{cases}$$

and $\nu(x) + \nu(1 - x) = 1$. (See Example 6 of Chapter 3.)

This can be put into the form of Proposition 10.3 by finding, for $w \in \left[\frac{2\pi}{3}, \frac{4\pi}{3}\right]$

$$h(w - \pi) = -\frac{d}{dw}\cos^2\left[\frac{\pi}{2}\nu\left(\frac{3}{2\pi}w - 1\right)\right]$$

$$= 2\cos\left[\frac{\pi}{2}\nu\left(\frac{3}{2\pi}w - 1\right)\right]\sin\left[\frac{\pi}{2}\nu\left(\frac{3}{2\pi}w - 1\right)\right]\frac{\pi}{2}\nu'\left(\frac{3}{2\pi}w - 1\right)\frac{3}{2\pi}.$$

If $\nu(x) \in C^1$ at least, then $\nu'\left(\frac{3}{2\pi}w - 1\right)$ has support in $\frac{2\pi}{3} \leq w \leq \frac{4\pi}{3}$ and hence $h(w)$ has support in $\left[-\frac{\pi}{3}, \frac{\pi}{3}\right]$. By a little algebraic manipulation we find

$$h(w) = \frac{3}{4} \sin\left[\pi\nu\left(\frac{3w}{2\pi} + \frac{1}{2}\right)\right] \nu'\left(\frac{3w}{2\pi} + \frac{1}{2}\right), \quad |w| \leq \frac{\pi}{3}.$$

The $\hat{\phi}(w)$ can be made to be in C^∞ by an appropriate $h(w)$.

This may be done by taking $e(x) = \exp(1/(x^2 - 1))\chi_{[-1,1]}(x)$. Then $h(w) = C_\epsilon e\left(\frac{w}{\epsilon}\right)$ satisfies the required conditions. This is the familiar regularization kernel of distribution theory. It may also be used to generate other $h(w)$, indeed let h_0 be any function ≥ 0 in L^1 with support on $[-\epsilon, \epsilon]$ and $\int h_0 = 1$. Then $h_1 = h * h_0$ has support on $[-2\epsilon, 2\epsilon]$ and belongs to C^∞. Since $\hat{\phi}^2(w) = (\chi_{[-\pi,\pi]} * h)(w)$, the appropriate new $\hat{\phi}$ may be given by

$$\hat{\phi}_1^2 = \chi_{[-\pi,\pi]} * h_1 = h_0^* \chi_{[-\pi,\pi]} * h$$

where $\hat{\phi}_0$ is the scaling function of h_0.

10.6 Dilations and other operations

We cannot expect V_0 to be invariant under all dilations. If it were, then all V_m would be the same as V_0 and $\bigcup_m V_m$ could not be dense in $L^2(\mathbb{R})$. Our definition of weak dilation invariance will therefore be

DEFINITION 10.2
Let $\{V_m\}$ be a multiresolution analysis of $L^2(\mathbb{R})$. Then $\{V_m\}$ is said to be <u>weakly dilation invariant</u> if for each $\alpha \in \mathbb{R}$ there is an $m \in \mathbb{Z}$ such that the dilation operator D_α maps V_0 into V_m.

Again the Haar wavelets and the ones based on splines are not weakly dilation invariant, but the Shannon wavelets are. This follows from the fact that $\widehat{D_\alpha\phi} = \frac{1}{\alpha}\hat{\phi}\left(\frac{w}{\alpha}\right) = \frac{1}{\alpha}\chi_{[-\alpha\pi,\alpha\pi]}(w)$ which is clearly a periodic multiple of $\hat{\phi}\left(\frac{w}{2}\right)$ for $1 < \alpha < 2$. Hence $D_\alpha\phi \in V_1$ for such α. Similar results hold for the Meyer type wavelets.

LEMMA 10.4
Let $\phi(t)$ be the scaling function of a Meyer type wavelet for some h_ϵ with support on $[-\epsilon, \epsilon]$, $\epsilon \leq \pi/3$. Then $D_\alpha\phi \in V_1$ if $\phi \in V_0$ for $1 \leq \alpha \leq 2(\pi - \epsilon)/(\pi + \epsilon)$.

The proof is straightforward since $\hat{\phi}\left(\frac{w}{\alpha}\right)$ has support in $[-\alpha(\pi + \epsilon), \alpha(\pi + \epsilon)]$. Thus $D_\alpha\phi \in V_1$ if $\hat{\phi}\left(\frac{w}{\alpha}\right) = m_\alpha\left(\frac{w}{2}\right)\hat{\phi}\left(\frac{w}{2}\right)$ where

$m_\alpha\left(\frac{w}{2}\right)$ is 4π periodic. This is possible if the support of $\hat\phi\left(\frac{w}{\alpha}\right)$ is contained in the set where $\hat\phi\left(\frac{w}{2}\right) = 1$, i.e., $[-2\pi+2\epsilon, 2\pi-2\epsilon]$. A translation of $\hat\phi\left(\frac{w}{\alpha}\right)$ by $2\pi k$, $k \neq 0$, will then have its support outside of the support of $\hat\phi\left(\frac{w}{2}\right)$. Hence the 4π periodic extension of $\hat\phi\left(\frac{2}{\alpha}\right)$ will give us $m_\alpha\left(\frac{w}{2}\right)$. In terms of α this is

$$\alpha(\pi + \epsilon) \leq 2\pi - 2\epsilon,$$

which gives us our conclusion.   □

COROLLARY 10.5
Let $\{V_m\}$ be the MRA of a Meyer type wavelet with $\epsilon < \frac{\pi}{3}$, then $\{V_m\}$ is weakly dilation invariant.

By the Lemma, if $\epsilon < \frac{\pi}{3}$, then $D_\alpha\phi \in V_1$ for $\alpha < \alpha_\epsilon$, where $\alpha_\epsilon = 2\left(\frac{\pi-\epsilon}{\pi+\epsilon}\right) > 2\frac{2\pi/3}{4\pi/3} = 1$. By a change of scale it follows that $D_\alpha\phi_1 \in V_2$ for $\phi_1 \in V_1$. Hence, for $\phi \in V_0$, $D_{2\alpha}\phi \in V_2$ and it follows that $D_{2^m\alpha}\phi \in V_{m+1}$, $m \geq 0$. Since $2^m\alpha$ must eventually be greater than 1, the conclusion follows.   □

We can combine the two results on dilation and translation to get

COROLLARY 10.6
Let $\{V_m\}$ be the MRA of a Meyer type wavelet $\psi(t)$ with $\epsilon < \frac{\pi}{3}$, let $\alpha, \beta \in \mathbb{R}$; then $\psi(\alpha t - \beta) \in V_m$ for some m.

Other operators associated with translation are difference operators, differential operators and convolution operators. The invariance of these operators can be based on Proposition 10.3, but stronger results can be obtained directly using the same techniques.

PROPOSITION 10.7
Let $\{V_m\}$ be the MRA of a Meyer type wavelet; let $P(\Delta)$ $(P(D))$ be a linear difference operator (differential operator) with constant coefficients; let H be the operator of convolution with $h \in L^1(\mathbb{R})$. Then for each $f \in V_m$, $P(\Delta)f$, $P(D)f$, and Hf belong to V_{m+1}.

The proof is along the same lines as the previous theorems and will be omitted.

10.7   Problems

1. Show directly that for the Shannon wavelets, the space V_m is invariant under differentiation.

2. Recall from Chapter 7 that the Green's function of the differential operator $D^2 y$ with boundary conditions $y(0) = y(1) = 0$ is

given by $g(x,t) = \begin{cases} x(1-t), & 0 \le x \le t \le 1 \\ t(1-x), & 0 \le t \le x \le 1 \\ 0, & \text{otherwise} \end{cases}$

and hence is in $L^1(\mathbb{R})$.

Show therefore that the solution to $D^2 y = f$, $f \in V_0$ satisfying the boundary conditions is also in V_0 for the Shannon wavelet case.

3. For the Franklin wavelets, show by example that neither translation invariance nor weak translation invariance holds for V_0.

4. Show directly that the weak translation invariant property holds for the Meyer wavelets generated by $h(\xi) = 1/2\epsilon$, $|\xi| < \epsilon \le \frac{\pi}{3}$, $h(\xi) = 0$, $\epsilon \le \xi$.

5. For the Franklin wavelets, show that the weak dilation invariance property does not hold.

11

Analytic Representations Via Orthogonal Series

The analytic representation of functions or distributions on the real line is usually given by a Cauchy type formula (Chapter 2, Section 4) [Br] but in some cases may also be given by an orthogonal series. This is evident for periodic functions and distributions for which trigonometric series may be used [W15]. For nonperiodic functions the Hermite series can be used, but it leads to analytic representations that are no longer of the same form [W16]. For functions with compact support, series of Legendre polynomials lead to analytic representations involving Legendre functions of the second kind [W17].

The more natural approach for arbitrary functions on $\mathbb{R}$ seems to be one involving wavelets [W18]. The analytic representation preserves dilations and translations. Since these are the operators used to define wavelets, their analytic representatives also are wavelets. This is also true for harmonic functions that may be obtained from analytic representations. The corresponding "harmonic wavelets" also satisfy the same defining relations.

In this chapter, we first present elements of the analytic representation theory involving the trigonometric, Hermite, and Legendre systems. We then consider the wavelet approach in more detail. Both functions and tempered distributions are considered since they lead to slightly different theories.

11.1 Trigonometric series

We have seen in Chapter 4 that each tempered distribution of period 2π has a Fourier series convergent in the sense of S'. The analytic representation of such distributions is most easily obtained by splitting up the Fourier series.

Let $f(x)$ be a periodic distribution of period 2π with Fourier series (see Chapter 4)

$$\sum_{n=0}^{\infty} a_n \cos nx + b_n \sin nx.$$

Consider the series

$$\sum_{n=0}^{\infty} (a_n - ib_n)e^{inz} \quad \text{and} \quad \sum_{n=0}^{\infty} (a_n + ib_n)e^{-inz}.$$

The first converges for all z in the upper half-plane, the second for all z in the lower half-plane since the a_n and b_n, as Fourier coefficients of a distribution, grow no faster than a power of n. Also both series converge in the sense of distributions for z on the real axis. Clearly their sum also does and, in fact, since it is just twice the Fourier series again, converges to $2f(x)$. Thus the function

$$f^+(z) = \sum_n \frac{a_n - ib_n}{2} e^{inz}$$

for Im $z > 0$,

$$f^-(z) = \sum_n \frac{-a_n - ib_n}{2} e^{-inz}$$

for Im $z < 0$ is an analytic representation of $f(x)$ since

$$\lim_{\epsilon \to 0} \{ f^+(x + i\epsilon) - f^-(x - i\epsilon) \}$$

$$= \lim_{\epsilon \to 0} \left\{ \sum_n \frac{a_n - ib_n}{2} e^{in(x+i\epsilon)} + \sum_n \frac{a_n + ib_n}{2} e^{-in(x-i\epsilon)} \right\}$$

$$= \lim_{\epsilon \to 0} \left\{ \sum_n (a_n \cos nx + b_n \sin nx) e^{-n\epsilon} \right\}$$

$$= \sum_n (a_n \cos nx + b_n \sin nx) \lim_{\epsilon \to 0} e^{-n\epsilon} = f(x).$$

Of course, the limits in the preceding computations must be taken in the sense of $\mathcal{S}'$ in which case the interchange of limits is justified.

These Fourier series of distributions are not convergent in the usual pointwise sense, but they may be pointwise summable at certain points. In fact we shall see that at points at which the distribution has a value in the sense of Lojasewicz [Lo] (see Section 8.2) the Fourier series is Abel summable to that value. We now recall (see Chapter 4) that the Abel means are

$$f(r,x) := \frac{1}{\pi} \int_{-\pi}^{\pi} f(x+t)P_r(t)dt,$$

where P_r, the Poisson kernel, is given by

$$P_r(t) = \frac{1-r^2}{2(1-2r\cos t + r^2)},$$

whence we may prove the statement about Abel summability by using the following.

LEMMA 11.1
Let $P_r(t)$ be the Poisson kernel. Then

$$K_r(t) = ((-t)^m P_r^{(m)}(t))/\pi m! \to \delta(t) \text{ as } r \to 1^- \text{ for } t \in (-\pi,\pi).$$

Moreover $K_r(t)$ is a quasi-positive type delta family.

PROOF It is well know that $P_r(t) \to \pi\delta(t)$ in the distribution sense on $(-\pi,\pi)$ as $r \to 1^-$. Both differentiation and multiplication by t are continuous operations with respect to distribution convergence. Hence

$$\frac{(-t)^m P_r^{(m)}(t)}{\pi m!} \to \frac{(-t)^m \delta^{(m)}(t)}{m!} = \delta(t) \quad \text{as } r \to 1^-.$$

The kernel $K_r(t)$ is a quasi-positive delta family if it satisfies

(i) $\displaystyle\int_{-\pi}^{\pi} K_r(t)dt = 1,$

(ii) $\displaystyle\int |K_r(t)|dt \le M$, a constant,   (11.1)

(iii) $K_r(t) \to 0$ uniformly on $[-\pi,-\delta]$ and $[\delta,\pi]$ for each $\delta > 0$.
(See Chapter 8 where a slightly more general definition is used.)

To prove (i) we use the fact that $(1/\pi)\int_{-\pi}^{\pi} P_r(t)dt = 1$ and integrate by parts m times. The integrated terms are all bounded and what we have left is

$$\frac{1}{\pi} \int_{-\pi}^{\pi} \frac{(-t)^m P_r^{(m)}(t)}{m!} dt.$$

To prove (ii) we use the inequality $|t^{m+1}P_r^{(m)}(t)| \le K_m$, a constant independent of r and t, which may be proved inductively. We then integrate

$$\int_{-\pi}^{\pi} |K_r(t)|\, dt = 2\int_0^{\pi} t^m |P_r^{(m)}(t)|\, dt$$

$$= \pm 2 \sum_{k=1}^{m+1} \int_{\tau_{k-1}}^{\tau_k} (-1)^k t^m P_r^{(m)}(t)dt$$

by parts m times. Here τ_k are the zeros of $P_r^{(m)}$ for $k = 1, \cdots, m$ and $\tau_0 = 0$, $\tau_{m+1} = \pi$.

The integrated terms are each of the form

$$\pm 2 \sum_k \left\{ \tau_k^{n+1} P_r^{(n)}(\tau_k) - \tau_{k-1}^{n+1} P_r^{(n)}(\tau_{k-1}) \right\}$$

which, by the preceding inequality, are bounded. The left-over integral is exactly

$$\pm 2 \sum_{k=1}^{m+1} \int_{\tau_{k-1}}^{\tau_k} (-1)^k P_r(t) dt,$$

which is bounded by the fact used in the proof of (i).

Finally, we prove (iii) by the inequality

$$|P_r^{(m)}(t)| \le \frac{A_m(1-r)}{t^{2m+2}},$$

which follows from the same inequality in the case $m = 0$ (see [Z, p.97]). □

THEOREM 11.2

The Fourier series of a distribution f with value γ at x_0 is Abel summable to γ at x_0.

PROOF To prove this we must show that $f(r, x_0) \to \gamma$ as $r \to 1^-$, where

$$f(r, x) = \frac{1}{\pi} \int_{-\pi}^{\pi} f(x+t) P_r(t) dt.$$

If we assume, as we may without loss of generality, that the Fourier series of f has constant terms 0, we may write $f = D^m F$, where F is a periodic and a continuous function. Moreover, by picking m large enough we may assume

$$\frac{F(x)}{(x - x_0)^m} \to \frac{\gamma}{m!} \quad \text{as} \quad x \to x_0.$$

The Abel means, by integrating by parts (the integrated terms are the 0-distribution), may be expressed as

$$f(r, x_0) = \frac{(-1)^m}{\pi} \int_{-\pi}^{\pi} F(t + x_0) P_r^{(m)}(t) dt$$

$$= \frac{m!}{\pi} \int_{-\pi}^{\pi} \frac{F(t + x_0)(-t)^m P_r^{(m)}(t)}{t^m \, m!} dt.$$

The first part of the lemma seems to suggest that the limit as $r \to 1^-$ should be

$$m! \int_{-\pi}^{\pi} \frac{F(t+x_0)}{t^m} \delta(t) dt = \lim_{t \to 0} m! \frac{F(t+x_0)}{t^m} = \gamma$$

and indeed would prove this statement if $F(t+x_0)/t^m$ were differentiable sufficiently often. Since it is not, we must use the fact that $K_r(t)$ is a kernel of quasi-positive type.

For such kernels $\int f K_r \to f(0)$ under the hypothesis that f is continuous at 0 (see Chapter 8). Hence our conclusion follows. □

The relation between the Abel means $f(r,x)$ and the analytic representation $f^{\pm}(z)$ of f is clear if we look at their respective series. In fact for $z = re^{ix}$ $f(r,x) = f^+(z) - f^-(\bar{z})$ so that any statement about Abel summability can be translated into a statement about analytic representation.

COROLLARY 11.3
Let f be a periodic distribution. If f has a value γ at x, then $f^+(x+i0) - f^-(x-i0) = \gamma$.

11.2 Hermite series

In the case of trigonometric Fourier series the analytic representation f^+ is distinguished by having its negative Fourier coefficients equal to zero. If we move to functions in $L^2(\mathbb{R})$, these analytic representations belong to the Hardy space $H^2(\mathbb{C}^+)$, but the Hermite functions do not. Hence the (f^+) cannot be characterized in terms of vanishing coefficients.

In order to determine which sequences of coefficients do belong to functions in $H^2(\mathbb{C}^+)$, we use the auxilliary sequences $\{\alpha_{nk}\}$ where

$$\alpha_{nk} := \int_0^\infty h_n h_k, \qquad n, k = 0, 1, \cdots .$$

Then $\sum i^n \alpha_{nk} h_n$ is the Hermite series of a function in $H^2(\mathbb{C}^+)$ since its Fourier transform vanishes on $(-\infty, 0)$. By using the properties of the Hermite functions we may derive the following properties of the α_{nk}:

$$\text{(i)} \quad \alpha_{nk} = \frac{1}{2}\delta_{nk}, \quad n+k \text{ even},$$

$$\text{(ii)} \quad \sum_{n=0}^\infty \alpha_{nk}\alpha_{nl} = \alpha_{kl}. \tag{11.2}$$

These furnish the basic tools to prove

PROPOSITION 11.4

Let $\sum c_n h_n$ be the Hermite series expansion of a function $f \in L^2(-\infty, \infty)$; let

$$a_n = \sum_{k=0}^{\infty} i^{n-k} \alpha_{nk} c_k, \quad n = 0, 1, \cdots,$$

$$\left(b_n = \sum_{k=0}^{\infty} (-i)^{n-k} \alpha_{nk} c_k, \quad n = 0, 1, \cdots \right);$$

then $\sum a_n h_n (\sum b_n h_n)$ is the expansion of a function in $H^2(C^+)$ $(H^2(C^-))$. Moreover $c_n = a_n + b_n$, and if f itself $\in H^2(C^+)$, $c_n = a_n$.

PROOF The proof involves using the infinite matrices $A = [i^{n-k} \alpha_{nk}]$ and $\overline{A} = [(-i)^{n-k} \alpha_{nk}]$ as operators on ℓ^2. They have the properties that $A + \overline{A} = I$ and $A\overline{A} = 0$. From these and the fact that A maps ℓ^2 into sequences that are the Hermite coefficients of H^2 functions, i.e., $H^2(C^+)$ or $H^2(C^-)$, the conclusions follow. □

Unfortunately, when we look at convergence of Hermite expansions of H^2 functions, we cannot in general extend the convergence on the real axis to the complex plane. In order to imitate the convergence of Fourier series of analytic representations, we need to introduce new functions associated with the Hermite functions. One such family of functions consists of projections of Hermite functions.

Since the Hermite functions themselves are not in H^2 and since Hermite series of H^2 functions do not behave very well off the real axis, it is better to consider other series instead. Thus, we shall consider the projections of the Hermite functions on H^2 and the behavior of series of the projections in the complex plane. These projections $\{g_k\}$ may be given by the functions

$$g_k(t) = \sum_{n=0}^{\infty} (i)^{n-k} \alpha_{nk} h_n(t), \quad k = 0, 1, 2, \cdots. \tag{11.3}$$

Then if $f \in H^2(C^+)$ with Hermite expansion $\sum a_n h_n$ we see that

$$f = \sum_n a_n h_n = \sum_n \left\{ \sum_k (i)^{n-k} \alpha_{nk} a_k \right\} h_n = \sum_n \left\{ \int f \overline{g_n} \right\} h_n$$

by Parseval's equality since by the symmetry of α_{nk}, the Hermite coefficients of $\overline{g_n}$ are $i^{n-k} \alpha_{nk}$. Thus we find that $a_n = \int f h_n = \int f \overline{g_n}$ for $f \in H^2(C^+)$ and similarly $\int f h_n = \int f g_n$ for $f \in H^2(C^-)$. Whenever $f \in L^2$ we can rewrite the results of Proposition 11.4 as

$$\int f h_n = \int f \overline{g_n} + \int f g_n,$$

whence it follows that

$$\overline{g_n} + g_n = h_n \quad a.e.$$

Thus all $f \in L^2$ may be written as

$$f = \sum c_n h_n = \sum c_n g_n + \sum c_n \overline{g_n}$$

where $\sum c_n g_n \in H^2(\mathbb{C}^+)$. These $\{g_n\}$ may be shown, by using the properties of the α_{nk}, to satisfy

$$g_k(x) = \frac{1}{\sqrt{2\pi}} \int_0^\infty e^{ixt}(-1)^k h_k(t)dt,$$

$$g_k(z) = \frac{1}{2\pi i} \int_{-\infty}^\infty \frac{h_k(x)}{x - z}dx, \quad \text{Im } z > 0.$$

Using these properties we may attack the problem of convergence in the complex plane.

THEOREM 11.5
Let $f \in L^2(-\infty, \infty)$ with Hermite series expansion $\sum c_n h_n$; let

$$f^+(z) = \frac{1}{2\pi i} \int_{-\infty}^\infty \frac{f(x)}{x - z}dx$$

be the projection of f on $H^2(\mathbb{C}^+)$, Im $z > 0$; then $\sum c_n g_n(z)$, g_n given by (11.3),
   (i)   *converges absolutely for Im $z > 0$,*
   (ii) *converges uniformly to $f^+(z)$ in any region for which Im $z \geq \epsilon > 0$,*
   (iii) *converges in the sense of $L^2(-\infty, \infty)$ for each fixed $y \geq 0$, and $\sum c_n \{\overline{g}_n(z) + g_n(z)\}$ converges to a harmonic function whose boundary value on $(-\infty, \infty)$ are given by $f(x)$.*

PROOF   The first conclusion follows from Bessel's inequality since the $g_n(z)$ are just the Hermite coefficients of $(2\pi i)^{-1}(x - z)^{-1}$. Hence we see that

$$\left| \sum_{n=0}^N c_n g_n(z) \right|^2 \leq \sum |c_n|^2 \sum |g_n(z)|^2 \leq \|f\|_2^2 \int_{-\infty}^\infty \frac{1}{4\pi^2|x - z|^2}dx.$$

To prove the uniform convergence we consider the equality

$$\left|\sum_{n=0}^{N} c_n g_n(z) - f^+(z)\right| = \frac{1}{2\pi}\left|\int_{-\infty}^{\infty} \frac{\sum_{n=0}^{N} c_n h_n(x) - f(x)}{x - z} dx\right|$$

$$\leq \left\|\sum_{n=0}^{N} c_n h_n - f\right\|_2 \left\{\frac{1}{2\sqrt{\pi y}}\right\}$$

from which our assertion follows immediately.

Finally to show L^2 convergence we use Plancherel's identity twice to calculate

$$\int_{-\infty}^{\infty} \left|\sum_{n=0}^{N} c_n g_n(x + iy) - f^+(x + iy)\right|^2 dx$$

$$= \int_{0}^{\infty} e^{-2yt}\left|\sum_{n=0}^{N} c_n h_n(t)(-i)^n - \hat{f}(t)\right|^2 dt$$

$$\leq \int_{-\infty}^{\infty} \left|\sum_{n=0}^{N} c_n h_n(t)(-i)^n - \hat{f}(t)\right|^2 dt$$

$$= \left\|\sum_{n=0}^{N} c_n h_n - f\right\|_2^2 \to 0 \text{ as } N \to \infty.$$

Here we have used the fact that

$$f^+(x + iy) = \frac{1}{2\pi i}\int_{0}^{\infty} e^{i(x+iy)t} \hat{f}(t) dt,$$

which is another consequence of Plancherel's identity. The last statement follows immediately from the others. □

We note that with the replacement of h_n by g_n the convergence of the series is improved off the real axis. However it is still not good enough to include a number of cases of interest. We know that each function of polynomial growth has a Hermite series with coefficients $O(n^p)$ convergent to it in the sense of tempered distributions. Also each such function has an analytic representation to which we should like the series with the g_n to converge with some sort of uniformity. Therefore, in order to improve the convergence we introduce another related sequence of functions $\{\tilde{h}_n\}$, the Hermite functions of the second kind. These are more closely analogous to the functions $\{e^{inz}\}$ used with Fourier series.

These Hermite functions of the second kind $\tilde{h}_n$ are solutions to the Hermite equation

$$(z^2 - D^2)w = (2n+1)w, \qquad n = 0, 1, \cdots,$$

which vanish at $i\infty$ when z is in the upper half-plane and at $-i\infty$ when z is in the lower half-plane. The functions were studied in [W16] where it is shown that they satisfy the same recurrence formulae as the Hermite functions themselves, as well as the following formulae:

$$\text{(i)} \quad \tilde{h}_n(z) = h_n(z) \int_{i\infty}^{z} \frac{2}{h_n^2}, \qquad \text{Im } z > 0,$$

$$\text{(ii)} \quad \tilde{h}_n(z)h_k(z) = -\int_{-\infty}^{\infty} \frac{h_n(x)h_k(x)}{x-z}dx, \qquad \text{Im } z \neq 0,\, n \geq k, \qquad (11.4)$$

$$\text{(iii)} \quad \tilde{h}_n(x+i0) - \tilde{h}_n(x-i0) = -2\pi i h_n(x).$$

THEOREM 11.6
Let f be a continuous real valued function of polynomial growth with Hermite series expansion $\sum c_n h_n$; then

$\quad$ (i) $\quad \sum (ic_n/2\pi)\tilde{h}_n(z)$ *converges uniformly in compact subsets of either the upper or lower open half-plane,*

$\quad$ (ii) $\quad \sum (ic_n/2\pi)\{\tilde{h}_n(z) - \tilde{h}_n(\bar{z})\}$ *converges in the upper half-plane to a real harmonic function $u(z)$,*

$\quad$ (iii) $\quad \lim_{y\to 0} u(x+iy) = f(x)$ *uniformly on compact sets of $\mathbb{R}$.*

PROOF   We have seen in Chapter 2 that a function of polynomial growth belongs to $\mathcal{S}'$, the space of tempered distributions; convergence in $\mathcal{S}'$ is equivalent to weak convergence, i.e., $\sum c_n h_n$ converges as $\sum c_n \langle h_n, \phi \rangle$ converges for each $\phi \in \mathcal{S}$. Also an element f of $\mathcal{S}'$ has a Hermite expansion $\sum c_n h_n$ whose coefficients $c_n = O(n^p)$ for some integer p, and which converges to f. Now by property (ii)

$$\sum \frac{ic_n}{2\pi}\tilde{h}_n(z) = \sum \frac{ic_n}{2\pi} \int_{-\infty}^{\infty} \frac{h_n(t)\exp(z^2/2)\exp(-t^2/2)}{z-t}dt$$

$$= \exp(z^2/2)\sum c_n \frac{1}{2\pi i}\int_{-\infty}^{\infty} \frac{h_n(t)}{t-z}\exp(-t^2/2)dt$$

$$= \exp(z^2/2)\sum c_n \langle \phi_z, h_n \rangle$$

where $\phi_z(t) = \exp(-t^2/2)/(t-z) \in \mathcal{S}$. Therefore this series converges pointwise for $\text{Im } z \neq 0$.

To show that this convergence is in fact uniform on compact sets we use the fact that $h_n = (x^2 - D^2)^{p+2}h_n/(2n+1)^{p+2}$, where p is the nonnegative integer associated with the asymptotic behavior of the coefficients. Then we have

$$\sum_n c_n \langle \phi_z, \frac{(x^2 - D^2)^{p+2}}{(2n+1)^{p+2}} h_n \rangle = \sum_n \frac{c_n}{(2n+1)^{p+1}} \langle (x^2 - D^2)^{p+2} \phi_z, h_n \rangle.$$

The series $\sum (c_n/(2n+1)^{p+2})$ converges absolutely and $\langle (x^2 - D^2)^{p+2} \phi_z, h_n \rangle$ is uniformly bounded on compact sets avoiding the real axis.

The convergence in (ii) follows from that in (i), and the limit function will clearly be harmonic. We need only show that it is real. This can be done by showing $i(\tilde{h}_n(z) - \tilde{h}_n(\bar{z}))$ to be real. But we have, by (11.4), with $k = 0$,

$$i(\tilde{h}_n(z) - \tilde{h}_n(\bar{z}))$$
$$= -i \int_{-\infty}^{\infty} h_n(t) \exp(-t^2/2) \left\{ \frac{\exp(z^2/2)}{t - z} - \frac{\exp(\bar{z}^2/2)}{t - \bar{z}} \right\} dt$$

whence, since the expression in brackets is purely imaginary, it follows that it is.

The last conclusion of Theorem 11.6 is a consequence of the fact that $u(z) \exp(-x^2/2)$ is the Poisson integral representation of $f(t) \exp(-t^2/2)$. Hence $u(z) \exp(-z^2/2) \to f(t) \exp(-t^2/2)$ uniformly on compact sets and hence our conclusion. $\square$

11.3 Legendre polynomial series

A function or distribution with support in $[-1, 1]$ may be given by a series of Legendre polynomials

$$f(x) = \sum_{n=0}^{\infty} a_n P_n(x) \chi(x)$$

where $\chi(x)$ is the characteristic function of $[-1, 1]$. The P_n, as we have seen in Chapter 6, satisfy the differential equation

$$((1 - x^2)y')' + n(n+1)y = 0.$$

Since this is a second order differential equation, it must have a second linearly independent solution Q_n which we take to be

$$Q_n(z) = 2^{-n-1} \int_{-1}^{1} (1 - t^2)^n (z - t)^{-n-1} dt, \quad \text{Im } z \neq 0$$

[Sz, p.74]. This solution is related to the analytic representation since

$$\frac{i}{\pi} Q_n(z) = \frac{1}{2\pi i} \int_{-1}^{1} P_n(t) \frac{1}{t - z} dt \quad \text{[Sz, p.75]}.$$

Thus it appears we have the following

PROPOSITION 11.7

Let $\{a_n\}$ be a sequence of numbers such that $a_n = O(n^p)$ for some integer $p \geq 0$; then the series $\sum a_n P_n \chi$ converges to a tempered distribution g and the series $\sum a_n \frac{i}{\pi} Q_n(z)$ converges to a holomorphic function $g^{\pm}(z)$ for Im $z \neq 0$, which is the analytic representation of g.

Details of the proof are omitted. They may be found in [W17].

11.4   Analytic and harmonic wavelets

The operator that maps a function or distribution into its analytic representation is linear and commutes with certain other operators. Indeed, if $A^{\pm}(f(x)) = f^{\pm}(z)$, then

$$\frac{d}{dz} f^{\pm}(z) = A^{\pm}\left(\frac{d}{dx} f(x)\right)(z).$$

The same is true for the dilation operator D_α and translation operator T_β, i.e.,

$$f^{\pm}(\alpha z) = A^{\pm}(D_\alpha f(x))(z)$$

and

$$f^{\pm}(z - \beta) = A^{\pm}(T_\beta f(x))(z).$$

This is obvious in the case of L^2 functions and can easily be shown for distributions by the "integration of parts" device.

The operator A^+ maps $L^2(\mathbb{R})$ into the Hardy space $H^2(\mathbb{C}^+)$ consisting of all functions in the upper half-plane $\mathbb{C}^+$ that satisfy

$$\sup_{y>0} \int_{-\infty}^{\infty} |f(x + iy)|^2 dx < \infty.$$

This is clear if we use the Fourier transform definition,

$$f^+(x + iy) = \frac{1}{2\pi} \int_0^{\infty} \hat{f}(\omega) e^{i\omega x} e^{-\omega y} d\omega,$$

since by Parseval's equality

$$\int_{-\infty}^{\infty} |f^+(x + iy)|^2 dx = \frac{1}{2\pi} \int_0^{\infty} |\hat{f}(\omega) e^{-\omega y}|^2 d\omega$$

$$\leq \frac{1}{2\pi} \int_0^{\infty} |\hat{f}(\omega)|^2 d\omega \leq \frac{1}{2\pi} \|f\|^2. \tag{11.5}$$

This inequality also shows that A^+ is a bounded operator whose range is all of $H^2(\mathbb{C}^+)$.

It is also possible to show that

    (i) $f^+(x+iy) \to f^+(x+i0)$ a.e.,

    (ii) $\|f^+(\cdot + iy) - f^+(\cdot + i0)\| \to 0$,

[S-W, p.90] so that $H^2(\mathbb{C}^+)$ may be considered a closed subspace of $L^2(\mathbb{R})$.

If we assume that the scaling function $\phi \in \mathcal{S}_r$ as usual, then it certainly is in $L^2(\mathbb{R})$ and has an associated analytic representation $\phi^\pm(z)$. The translations of ϕ map into those of ϕ^+, and the subspace V_0 maps into a closed subspace V_0^+ of $H^2(\mathbb{C}^+)$.

Since $\{\phi(x-n)\}$ is an orthonormal basis of V_0, it follows by (11.5) that

$$\|f^+(z) - \Sigma a_n \phi^+(z-n)\|^2 \le \frac{1}{2\pi}\|f - \Sigma a_n \phi(\cdot - n)\|^2, \qquad (11.6)$$

and hence V_0^+ is the closed linear span of $\{\phi^+(z-n)\}$. Unfortunately $\{\phi^+(z-n)\}$ is not necessarily orthogonal, since for $n \neq 0$

$$\int_{-\infty}^{\infty} \phi^+(x+i0-n)\overline{\phi^+(x+i0)}dx = \frac{1}{2\pi}\int_0^\infty |\hat{\phi}(\omega)|^2 e^{-i\omega n}d\omega \neq 0,$$

but it is a <u>basis</u>. Indeed, we need merely show the uniqueness of the expansion $\Sigma a_n \phi^+(z-n)$ to conclude this. This in turn follows if $\hat{\phi}(\omega) \neq 0$ a.e. on an interval of length 2π in $[0,\infty)$ since if

$$\sum_n a_n \phi^+(z-n) = 0,$$

then

$$\sum_n a_n e^{i\omega n}\hat{\phi}(\omega)\chi_{[0,\infty)}(\omega) = 0.$$

Thus it follows that $\Sigma a_n e^{i\omega n} = 0$ a.e. in some interval of length 2π. Hence by the uniqueness theorem for Fourier series $a_n = 0$, $n \in \mathbb{Z}$.

The dilation subspaces V_m also map onto subspaces V_m^+ of $H^2(\mathbb{C}^+)$, which are dilations of V_0^+. This gives

PROPOSITION 11.8
Let $\phi \in \mathcal{S}_r$ be a scaling function such that $\hat{\phi}(\omega) \neq 0$ a.e. on an interval of length 2π in $[0,\infty)$ with associated multiresolution analysis $\{V_m\}$ of $L^2(\mathbb{R})$; then $V_m^+ = A^+ V_m$, $m \in \mathbb{Z}$ satisfies

$$\cdots \subset V_{-1}^+ \subset V_0^+ \subset V_1^+ \subset \cdots \subset V_m^+ \subset \cdots \subset H^2(\mathbb{C}^+),$$

$$\cap_m V_m^+ = \{0\}, \qquad \overline{\cup_m V_m^+} = H^2(\mathbb{C}^+),$$

and $\{\phi^+(2^m x - n))$ *is a basis of* V_m^+.

The same thing can be done for the wavelets themselves; the analytic representation of the mother wavelet is

$$\psi^\pm(z) = \frac{1}{2\pi i} \int_{-\infty}^{\infty} \frac{\psi(x)}{x - z} dx$$

and the <u>analytic wavelets</u> $\psi_{mn}^\pm$ are obtained by dilation and translation of $\psi^\pm$. The subspaces W_m with orthogonal basis $\{2^{m/2}\psi(2^m x - n)\}$ is mapped onto W_m^+ with basis $\{\psi^+(2^m z - n)\}$. Moreover $W_m^+ \cap V_m^+ = \{0\}$. Suppose, on the contrary for $m = 0$, that $g^+ \in W_0^+ \cap V_0^+ \neq 0$. Then

$$\hat{g}(\omega) = \sum_n a_n e^{in\omega} \hat{\psi}(\omega) = \sum_n b_n e^{in\omega} \hat{\phi}(\omega) \text{ a.e., } \omega > 0 \qquad (11.7)$$

for some $\{a_n\}$ and $\{b_n\}$ not zero. Since

$$\hat{\phi}(\omega) = m_0\left(\frac{\omega}{2}\right)\hat{\phi}\left(\frac{\omega}{2}\right),$$

and

$$\hat{\psi}(\omega) = e^{-i\omega/2}\overline{m_0\left(\frac{\omega}{2} + \pi\right)}\hat{\phi}\left(\frac{\omega}{2}\right),$$

it follows that in some interval of length 4π,

$$\alpha(\omega)e^{-i\omega/2}\overline{m_0\left(\frac{\omega}{2} + \pi\right)} = \beta(\omega)m_0\left(\frac{\omega}{2}\right), \text{ a.e.} \qquad (11.8)$$

(where $\alpha(\omega) = \Sigma a_n e^{i\omega n}$, $\beta(\omega) = \Sigma b_n e^{i\omega n}$). But both sides of (11.8) are periodic (4π) and hence equal a.e. in $\mathbb{R}$. Thus (11.7) holds a.e. in $\mathbb{R}$ and its inverse Fourier transform $g(x) \in W_0 \cap V_0$ and must be zero.

The same calculation works for V_m^+, and hence we may express V_{m+1}^+ as the nonorthogonal direct sum

$$V_{m+1}^+ = V_m^+ \oplus W_m^+.$$

This enables us to rewrite one conclusion of Proposition 11.8 as

$$\bigoplus_m W_m^+ = H^2(\mathbb{C}^+).$$

By the same argument as in V_m^+, $\{\psi^+(2^m x - n)\}_{n \in \mathbb{Z}}$ is a basis of W_m^+, but $\{\psi_{mn}^+\}$ is not a basis of all of $H^2(\mathbb{C}^+)$.

Indeed let $f \in L^2(\mathbb{R})$ be such that $\hat{f}(\omega) = 0$, $\omega > 0$ but $\|f\| \neq 0$. Then

$$f = \Sigma b_{mn}\psi_{mn}$$

where not all b_{mn} are zero. But

$$0 = f^+ = \Sigma b_{mn} \psi_{mn}^+,$$

and hence we do not obtain the uniqueness needed.

We can, by an identical procedure, obtain a multiresolution analysis of $H^2(\mathbb{C}^-)$, the Hardy space of analytic functions in the lower half-plane. Moreover we can express each $f \in L^2(\mathbb{R})$ as a linear combination of elements of $H^2(\mathbb{C}^+)$ and $H^2(\mathbb{C}^-)$, which are orthogonal, i.e.,

$$L^2(\mathbb{R}) = H^2(\mathbb{C}^+) \bigoplus H^2(\mathbb{C}^-) = \bigoplus_m (W_m^+ \oplus W_{\overline{m}}).$$

By taking $f^+(z) - f^-(\overline{z})$ we are able to find the harmonic function in the upper half-plane with boundary values given by the L^2 function f. It can be expressed in terms of the analytic wavelet expansion as

$$u(x, y) = \sum_{mn} b_{mn} [\psi_{mn}^+(x + iy) - \psi_{mn}^-(x - iy)]$$

$$= \sum_{n=-\infty}^{\infty} a_{mn} [\phi_{mn}^+(x + iy) - \phi_{mn}^-(x - iy)]$$

$$+ \sum_{k=m}^{\infty} \sum_{n=-\infty}^{\infty} b_{kn} [\psi_{kn}^+(x + iy) - \psi_{kn}^-(x - iy)].$$

Thus one of the attractive features of wavelet theory, the decomposition algorithm, is present here also. The finest scale of interest is chosen and the series

$$u_m(x, y) = \sum_n a_{mn} [\phi_{mn}^+(x + iy) - \phi_{mn}^-(x - iy)]$$

is used to approximate $u(x, y)$. The error in going to the next finer scale is given by the wavelet terms

$$\sum_n b_{m-1,n} [\psi_{m-1,n}^+(x + iy) - \psi_{m-1,n}^-(x - iy)].$$

Since the a_{mn} and b_{mn} are the same as before they are still related by (3.20).

11.5 Analytic solutions to dilation equations

We have seen in Chapter 5 that dilation equation (3.3) with rapidly decreasing coefficients

$$\phi(t) = \sqrt{2} \sum_k c_k \phi(2t - k)$$

always has a distribution solution in $\mathcal{S}'$. By taking the analytic representation of the solution we obtain a function $\phi^+(z)$, which satisfies (3.3) in the upper half-plane, and similarly in the lower half-plane.

By analogy to the last section we can define subspaces T_m^+ (T_m^-) consisting of analytic functions in the upper (lower) half-plane. For example, we have

$$T_0^+ := \left\{ f(z) = \sum_n a_n \phi^+(z - n) \mid a_n = O(|n|^p) \right\}.$$

However, we cannot do much with this since we cannot characterize $\overline{\cup_m T_m}$ (Section 5.2) except that it is a subspace of $\mathcal{S}'$.

Example. Let $c_0 = \sqrt{2}$, $c_k = 0$, $k \neq 0$; then the solution to the dilation equation in $\mathcal{S}'$ is $\phi(t) = \delta(t)$.

The corresponding analytic solution in the upper half-plane is $\phi^+(z) = -\frac{1}{2\pi i z}$. If $f \in T_m$, i.e., if $f = \Sigma a_n \delta(2^m t - n)$ then

$$f^+(z) = -\frac{1}{2\pi i} \sum_n a_n (2^m z - n)^{-1}.$$

The dilation equation (3.3) also has a solution composed of harmonic functions $u(x, y)$ in the upper half-plane, i.e.,

$$u(x, y) = \sqrt{2} \sum_k c_k u(2x - k, y).$$

This is again given by $u(x, y) = \phi^+(x + iy) - \phi^-(x - iy)$.

The construction of $\phi^+(x)$ and $u(x, y)$ can proceed in two ways. One is first to construct the solution in $\mathcal{S}'$ by starting with $\phi_0(t) = \delta(t)$ and letting

$$\phi_n(t) = \sqrt{2}\Sigma c_k \phi_{n-1}(2t - k), \qquad n = 1, 2, \cdots \qquad (11.9)$$

and then finding the analytic representation of the limit in $\mathcal{S}'$. The other is to start with $\phi_0^+(z) = -1/2\pi i z$ and use (11.9) to find $\phi_n^+(z)$ directly and then taking the pointwise limit in the upper half-plane. The latter is more amenable to computation.

11.6 Analytic representation of distributions by wavelets

In order to represent an element of $\mathcal{S}'$ by series of analytic and harmonic wavelets, we impose conditions on the scaling function again and assume $\phi \in \mathcal{S}_r$. The spaces T_m^+ and T_m^- are now composed of analytic functions in the upper and lower half-plane respectively whose boundary functions are continuous functions of polynomial growth (Chapter 5).

Since $\overline{UT_m} = S'_r$ we might expect to obtain analytic representations in terms of wavelets for each $f \in S'_r$,

$$f^+(z) = \sum_{n=-\infty}^{\infty} a_n \phi^+(z-n)$$

$$+ \sum_{m=0}^{\infty} \sum_{n=-\infty}^{\infty} b_{mn} 2^{m/2} \psi^+(2^m z - n),$$

but the first series may not converge so we proceed differently.

The analytic representation represents a continuous map from S'_r to a corresponding space of analytic functions, i.e., if $f_m \to f$ in S'_r then $f_m^+(z) \to f^+(z)$ uniformly on bounded subsets of the upper half-plane. This of course requires that the proper analytic representation be chosen. One such choice is as follows: any $f \in S'_r$ (or S') may be given as a finite order derivative of a continuous function of polynomial growth, say $F(x)$

$$f = D^k F.$$

An analytic representation of F is given by

$$F^+(z) = \frac{1}{2\pi i} \int_{-\infty}^{\infty} \frac{F(x)}{x - z} \left(\frac{z^2 + 1}{x^2 + 1} \right)^p dx$$

where p is sufficiently large to assure that $F(x)/(x^2+1)^p \in L^2(\mathbb{R})$. Then $f^+(z)$ is defined by

$$f^+(z) = D_z^k F^+(z).$$

The approximation to f from the space T_m is the function

$$f_m(x) = (f, q_m(\cdot, x)).$$

Thus if the analytic representation is taken as above

$$f_m^+(z) \to f^+(z).$$

We may also express f_m as

$$f_m = f_0 + f_m - f_0 = f_0 + \sum_{k=0}^{m-1} \sum_{n=-\infty}^{\infty} b_{kn} \psi_{kn},$$

and if the inner series converges we have

$$f_m^+(z) - f_0^+(z) = \sum_{k=0}^{m-1} \sum_{n=-\infty}^{\infty} b_{kn} \psi_{kn}^+(z) + g_m(z) \tag{11.10}$$

where $g_m(z)$ is an entire function.

LEMMA 11.9
Let $b_n = O(|n|^{r-\epsilon})$, $\epsilon > 0$; then

$$\sum_{n=-\infty}^{\infty} b_n \psi^+(z - n)$$

converges uniformly on compact subsets of the open upper half-plane.

PROOF   The proof is based on the moment property (Chapter 3)

$$\int_{-\infty}^{\infty} x^p \psi(x) dx = 0, \qquad p = 0, 1, \cdots, r.$$

Then

$$
\begin{aligned}
z^p \psi^+(z) &= \frac{1}{2\pi i} \int_{-\infty}^{\infty} \frac{z^p - x^p}{x - z} \psi(x) dx + \frac{1}{2\pi i} \int_{-\infty}^{\infty} \frac{x^p \psi(x)}{x - z} dx \\
&= \frac{1}{2\pi i} \int_{-\infty}^{\infty} \frac{x^p \psi(x)}{x - z} dx
\end{aligned}
\tag{11.11}
$$

holds for $p \leq r + 1$. Hence $|z^{r+1}\psi^+(z)|$ is uniformly bounded for $\operatorname{Im} z \geq \epsilon > 0$, and the conclusion follows. Thus if the coefficients in (11.10) satisfy the hypothesis, the series converges. $\square$

LEMMA 11.10
Let $f \in S_p'$, $\phi \in S_r$, $p < r$; then the wavelet coefficients of f satisfy

$$|b_{mn}| = |(f, \psi_{mn})| \leq C(|n| + 2^m)^{p+\frac{1}{2}}. \tag{11.12}$$

PROOF   We use the fact that each $f \in S_p'$ is of the form

$$f = D^p[(x^2 + 1)^j \mu]$$

for some integer j and finite measure μ on $\mathbb{R}$ [Ze, p.109].
    Also $\psi(x)$ satisfies the inequality

$$|\psi^{(k)}(x)| \leq \frac{C_j}{(1 + |x|)^j}, \qquad k = 0, 1, \cdots, r, \ j \geq 0,$$

since $\psi \in S_r$. Hence after integrating by parts p times we obtain

$$
\begin{aligned}
|b_{mn}| &\leq \int_{-\infty}^{\infty} (1 + x^2)^j |\psi_{mn}^{(p)}(x)| d|\mu| \\
&\leq \int_{-\infty}^{\infty} (1 + |x|)^{2j} \frac{C_{2j} 2^{m/2 + pm}}{(1 + |2^m x - n|)^{2j}} d|\mu| \\
&\leq (1 + |n|2^{-m})^{2j} C_{2j} 2^{m/2 + pm} \int_{-\infty}^{\infty} d|\mu|.
\end{aligned}
$$

The inequality (11.12) follows by taking $2j = p + \frac{1}{2}$. $\square$

We now combine the two lemmas to obtain the convergence we need.

LEMMA 11.11
Let $f \in S'_p$, $\phi \in S_r$, $p < r$, $b_{mn} = (f, \psi_{mn})$; then the series

$$\sum_{m=0}^{\infty} \sum_{n=-\infty}^{\infty} b_{mn} \psi_{mn}^+(z)$$

converges uniformly on the half-plane Im $z \geq 1$.

PROOF That the inner series converges is immediate, but we have to be a little more careful with the outer series. By (11.11) we find that

$$|(2^m z - n)^{r+1} \psi_{mn}^+(z)| \leq \frac{1}{2\pi} \int \frac{|x^{r+1} \psi(x)| 2^{m/2} dx}{|x - 2^m z + n|}$$

$$\leq \frac{1}{2\pi} 2^{-m/2} \int |x^{r+1} \psi(x)| dx$$

for Im $z \geq 1$. Hence the inner series satisfies

$$\sum_n |b_{mn} \psi_{mn}^+(z)| \leq C' \sum_n \frac{(|n| + 2^m)^{p+\frac{1}{2}}}{|2^m z - n|^{r+1}} 2^{-m/2}$$

$$\leq C' \sum_n \left\{ \frac{(|n| + 2^m)^2}{(n - 2^m \operatorname{Re} z)^2 + 2^{2m}} \right\}^{\frac{p}{2} + \frac{1}{4}} \cdot$$

$$2^{-m/2} [(n - 2^m \operatorname{Re} z)^2 + 2^{2m}]^{\frac{p}{2} - \frac{r}{2} - \frac{1}{4}}$$

since $|2^m z - n|^2 = (n - 2^m \operatorname{Re} z)^2 + (2^m \operatorname{Im} z)^2$.

In order to get a bound on the expression in parentheses we observe that

$$\frac{(a+1)^2}{(a-b)^2 + 1} \leq (b+1)^2 + 1 \qquad \text{for } a, b > 0.$$

By taking $a = |n| 2^{-m}$ and $b = \pm \operatorname{Re} z$ we obtain a bound that depends only on z. The series is thus dominated by the remaining factor

$$2^{-m/2} \sum_{n=-\infty}^{\infty} [(n - 2^m \operatorname{Re} z)^2 + 2^{2m}]^{\frac{p}{2} - \frac{r}{2} - \frac{1}{4}}$$

$$\leq 2^{-m/2} C'' \sum_{n=-\infty}^{\infty} (n^2 + 2^{2m})^{-3/4} \leq C''' 2^{-m/2}.$$

Thus the terms in the outer series are dominated by a geometric series and converge uniformly on the half-plane. $\square$

We now return to the analytic representation. By taking the limit in (11.10) as $m \to \infty$, we find

$$f^+(z) = f_0^+(z) + \sum_{k=0}^{\infty} \sum_{n=\infty}^{\infty} b_{kn} \psi_{kn}^+(z) + g_\infty(z), \qquad (11.13)$$

where $g_\infty(z) = \lim_{m \to \infty} g_m(z)$ is also an entire function. We can drop this g_∞ in (11.13) since an analytic representation plus an entire function is again an analytic representation.

THEOREM 11.12
Let $f \in S_p'$, $\phi \in S_r$, $p < r$, and let $b_{mn} = \langle f, \psi_{mn} \rangle$, $m = 0, 1, 2, \cdots$, $n = 0, \pm 1, \pm 2, \cdots$, be the wavelet coefficients of f. Then an analytic representation of f is given by

$$f^\pm(z) = f_0^\pm(z) + \sum_{m=0}^{\infty} \sum_{n=-\infty}^{\infty} b_{mn} \psi_{mn}^\pm(z) \qquad (11.14)$$

where the series converges for Im $z \geq 1(+)$ or Im $z \leq -1(-)$, where $f_0^\pm(z)$ is an analytic representation of f_0, the projection of f on T_0.

We have only worked out the convergence for f^+ but that for f^- is parallel. Also we have only shown the convergence for $|\text{Im } z| \geq 1$, but we may continue the sum of the series analytically to the upper (lower) half-plane.

11.7   Problems

1. Find the Fourier series of $\delta'^*(x)$ (the periodic extension of $\delta'(x)$), and use it to find its analytic representation.

2. Show that the analytic representation in Problem 1 satisfies the conclusion of Corollary 11.3 for $x \neq 0$, $x \in (-\pi, \pi)$, $\gamma = 0$.

3. Find the analytic representation as well as the Hermite series of $\delta'(x)$, $x \in \mathbb{R}$.

4. Show that for Problem 3 the series for a_n, $\sum_{k=0}^{\infty} i^{(n-k)} c_k \alpha_{nk}$ diverges for $n = 0$. (The exact values of c_k and α_{ok} may be found by using the recurrence formulae for the Hermite functions. For α_{ok} first use $\int_0^{\infty} (2k+1) h_k h_o = \int_0^{\infty} (x^2 - D^2) h_k h_o$ and integrate by parts twice.)

5. Find the analytic representation $\phi^{\pm}(z)$ of the scaling function of the Haar wavelet and show directly that $\phi^{\pm}(z-n)$ is the analytic representation of $\phi(x-n)$.

6. Find the analytic representation $\phi^{\pm}(z)$ of the scaling function of the Shannon wavelet by using its Fourier transform. Find a sequence $\{a_n\} \neq 0$ such that $\sum_n a_n \phi^+(z-n) = 0$. (Use the Fourier coefficients of a function that has support in $(-\pi, 0)$.

7. Repeat Problem 6 for the Meyer wavelets (except you will not find a closed form for $\phi^{\pm}$).

8. Show that the series

$$\sum a_n \phi^+(z-n)$$

diverges when ϕ is the Shannon scaling function, and a_n are the coefficients of $F(x) = x$.

12

Orthogonal Series in Statistics

Orthogonal series form an important tool in many branches of statistics. They are used in probability density estimation, in spectral density estimation of time series, in linear and nonparametric regression, and in Bayes and empirical Bayes estimation, among others.

We can only present a sampling of these applications. We choose several for which wavelets can also be used. We shall concentrate on methods used in probability density estimation and only treat the others briefly.

The problem is to try to estimate properties of a probability density function $f(x)$ from an independent sample $X_1, X_2, \cdots X_N$ drawn from a distribution with this density. For example, the mean and variance of the sample may be used to estimate the mean and variance of $f(x)$. If the sample is large enough, the density itself can be estimated. One way of doing so is by means of the empiric distribution

$$f^*(x) := \frac{1}{N} \sum_{i=1}^{N} \delta(x - X_i). \tag{12.1}$$

This is an unbiased estimator for continuous $f(x)$. This means that the expected value of $f^*(x)$ is $f(x)$, which is immediate from the properties of expected values,

$$E(f^*(x)) = \frac{1}{N} \sum_{i=1}^{N} E(\delta(x - X_i))$$

$$= \frac{1}{N} \sum_{i=1}^{N} \int_{-\infty}^{\infty} \delta(x - y_i) f(y_i) dy_i = \frac{1}{N} \sum_{i=1}^{N} f(x) = f(x).$$

The indefinite integral of $f^*(x)$ is the empiric distribution function $F^*(x)$. By the law of the iterated logarithm $F^*(x) \to F(x)$, the distribution function given by the integral of $f(x)$, almost surely (almost everywhere with respect to the measure induced by $f(x)$) [Ki]. The rate of convergence is $O\left(\left(\frac{\log\log N}{N}\right)^{1/2}\right)$.

Unfortunately the estimator (12.1) is very irregular even when $f(x)$ is very smooth. In order to obtain a smooth estimator a number of different procedures have been introduced (and widely studied). The simplest is the histogram estimator composed of piecewise constant functions. It's more regular than $f^*(x)$ but still not smooth; in terms of wavelet subspaces it is defined as

$$\tilde{f}_m(x) := \frac{1}{N} \sum_{i=1}^{N} q_m(X_i, x) \qquad (12.2)$$

where $q_m(x, y)$ is the reproducing kernel of V_m in prototype 1. The scaling function $\phi(x) = \chi_{[0,1)}(x)$, and $q(x,y) = \phi(x - [y])$, in this case. Since $\phi(2^m X_i - [2^m x])$ counts the number of X_i's in the interval $[k2^{-m}, (k+1)2^{-m})$ containing x, this is the same as the usual definition of histogram with nodes at $2^{-m}k$.

Other traditional smoother estimators are based on kernels [P], on Fourier series [K-T], on splines [Wh], and on general delta-sequences [W-B]. Some estimators converge in the mean square sense for all continuous densities, and some converge more rapidly for C^∞ densities, but none do both. Smooth estimators based on wavelets do both. This comes out of the properties of delta sequences of wavelets discussed in Chapter 8.

We first explore the properties of a Fourier series and a Hermite series estimator and then come back to wavelet estimators.

12.1 Fourier series density estimators

The Fourier series estimator of a continuous density function $f(x)$ is obtained by approximating the delta functions in (12.1) by the delta sequence consisting of the Dirichlet kernel

$$D_m(t) = \frac{\sin\left(m + \frac{1}{2}\right)t}{2\pi \sin\frac{1}{2}t} \qquad (12.3)$$

or by the Fejer kernel

$$F_m(t) = \frac{\sin^2(m+1)t/2}{2\pi(m+1)\sin^2 t/2} \qquad (12.4)$$

We have seen in Chapter 4 that both, when multiplied by $\chi_{[-\pi,\pi]}(t)$, are delta sequences and that $\{F_m\}$ is a positive delta sequence. The sorts of densities that can be estimated by using them are ones with support in $[-\pi, \pi]$. The associated estimators are

$$\tilde{f}_m^{(1)}(x) := \frac{1}{N} \sum_{i=1}^{N} D_m(x - X_i)$$

$$= \sum_{n=-m}^{m} \tilde{c}_n e^{inx} \qquad |x| \le \pi, \; m = 0, 1, \cdots \qquad (12.5)$$

where

$$\tilde{c}_n = \frac{1}{2\pi N} \sum_{i=1}^{N} e^{-inX_i}, \quad n = 0, \pm 1, \pm 2, \cdots,$$

and

$$\tilde{f}_m^{(2)}(x) := \frac{1}{N} \sum_{i=1}^{N} F_m(x - X_i),$$

$$= \sum_{n=-m}^{m} \tilde{c}_n \left(1 - \frac{|n|}{2m+1}\right) e^{inx}, \quad |x| \le \pi, \; m = 0, 1, \cdots. \qquad (12.6)$$

We would like the estimator of the density $f(x)$ to converge to it in some sense as the sample size N gets larger and as $m \to \infty$. One such sense is mean square convergence given by the requirement that

$$E[\tilde{f}_m(x) - f(x)]^2 \to 0.$$

This expression may be split up into a variance and a bias part

$$E[\tilde{f}_m(x) - f(x)]^2 = E[\tilde{f}_m(x) - f_m(x)]^2 + [f_m(x) - f(x)]^2 \qquad (12.7)$$

$$= \nu_m(x) \quad + \beta_m^2(x)$$

where $f_m(x) = E(\tilde{f}_m(x))$. We have already considered the bias part in Chapter 4. This is just the difference between $f(x)$ and the partial sums (or the Cesàro means) of its Fourier series.

This is clear from (12.5) since

$$E(\tilde{c}_n) = \frac{1}{2\pi N} \sum_{i=1}^{N} \int_{-\pi}^{\pi} e^{-inx} f(x) dx = \frac{1}{N} \sum_{i=1}^{N} c_n = c_n.$$

$$\nu_m(x) = E\left[\tilde{f}_m^2(x) - f_m^2(x)\right]$$

$$= \frac{1}{N} \left[\int_{-\pi}^{\pi} \delta_m^2(x, t) f(t) dt - \left(\int_{-\pi}^{\pi} \delta_m(x, t) f(t) dt\right)^2\right] \qquad (12.8)$$

$$\le \frac{1}{N} \|f\|_\infty \, \|\delta_m(x, \cdot)\|_2^2,$$

where δ_m is either (12.3) or (12.4). In both cases $\|\delta_m(x,\cdot)\|_2^2 = 0(m)$, which gives a bound

$$\nu_m(x) \leq \frac{c_1 m}{N}, \qquad m, N = 1, 2, \cdots \tag{12.9}$$

Since $f(x)$, being a density, cannot be periodic while its Fourier series is, we cannot expect the bias $\beta_m(x)$ to converge to zero very rapidly unless the periodic extension of $f(x)$ belongs to the Sobolev space $H_{2\pi}^\alpha$. This will happen, e.g., if the original $f \in C^\alpha$ and $f^{(k)}(\pm\pi) = 0$, $k = 0, 1, \cdots, \alpha - 1$.

PROPOSITION 12.1

Let f be a continuous density function with support on $[-\pi, \pi]$, then

(i) $\quad E\left|\tilde{f}_m^{(2)}(x) - f(x)\right|^2 \rightarrow 0$ *uniformly as $N, m \rightarrow \infty$, $\frac{m}{N} \rightarrow 0$;*

if $f \in C^\alpha$, $\alpha > 1$ and $f^{(k)}(\pm\pi) = 0$, $k = 0, 1, \cdots, \alpha - 1$, then

(ii) $\quad E\left|\tilde{f}_m^{(1)}(x) - f(x)\right|^2 = O(N^{-1+1/(2\alpha-1)})$, $m = \left[N^{\frac{1}{2\alpha-1}}\right]$

(iii) $\quad E\left|\tilde{f}_m^{(2)}(x) - f(x)\right|^2 = O(N^{-2/3})$, $m = [N^{1/3}]$.

PROOF For (i) we need merely invoke the uniform convergence of the Cesàro means (Chapter 4) to deduce that $\beta_m^2(x) \rightarrow 0$ uniformly. The same is true for $\nu_m(x)$ by (12.9).

For (ii) we use the crude estimate $|c_n n^\alpha| \leq c$ and hence

$$|\beta_m(x)| = \left|\sum_{|n|>m} c_n e^{inx}\right| \leq \sum_{|n|>m} \frac{c}{n^\alpha} = O(m^{1-\alpha}).$$

Thus by taking $m = [N^{1/(2\alpha-1)}]$ the bias and variance terms are balanced and the mean square error is

$$O(m^{2-2\alpha}) = O(m/N) = O(N^{(2-2\alpha)/(2\alpha-1)}).$$

In the case of (iii) the rate of convergence of the bias is $O(m^{-1})$ no matter how smooth f is [Z, p.122]. The error is given by the same calculation as in (ii). $\square$

REMARK If we only know that f is continuous, then $E|\tilde{f}_m^{(1)}(x) - f(x)|^2$ may not converge to zero. This is because the bias does not converge to zero for some continuous function as remarked in Chapter 4. In this case, $\tilde{f}_m^{(2)}$ is the better estimator. On the other hand, if $f \in C^\infty$ the convergence rate for $\tilde{f}_m^{(1)}$ approaches $O(N^{-1})$ while for $\tilde{f}_m^{(2)}$ is never better than $O(N^{-2/3})$.

12.2   Hermite series density estimators

Fourier series estimators because of their periodicity are appropriate only when the density has compact support. In general it does not, and a method that gives an estimator on the entire real line is preferred. One such method is based on the Hermite functions considered in Chapter 6, $\{h_n\}$ [Sch], [W14]. They constitute a complete orthogonal system in $L^2(\mathbb{R})$.

The estimator is similar to the Fourier series estimator; it is

$$\tilde{f}_m(x) = \sum_{n=0}^{m} \tilde{a}_n h_n(x) \tag{12.10}$$

where

$$\tilde{a}_n = \frac{1}{N} \sum_{i=1}^{N} h_n(X_i).$$

In order to find the rate of convergence we use the fact (Chapter 6) that

$$|h_n(x)| \leq C(n+1)^{-1/12}, \qquad n = 0, 1, \cdots.$$

The coefficients are seen to satisfy

$$E(\tilde{a}_n - a_n)^2 = \frac{1}{N^2} \sum_{i=1}^{N} E(h_n^2(X_i))$$

$$+ \frac{1}{N^2} \sum_{i=1}^{N} \sum_{j \neq i} E(h_n(X_i)h_n(X_j))$$

$$- \frac{2}{N} \sum_{i=1}^{N} E(h_n(X_i))a_n + a_n^2 \tag{12.11}$$

$$= \frac{1}{N^2} \sum_{i=1}^{N} E(h_n^2(X_i)) - a_n^2 \leq \frac{C^2}{N}(n+1)^{-1/6}$$

since $a_n = \int_{-\infty}^{\infty} f(x)h_n(x)dx = E(h_n(X_i))$.

The mean square error of this estimator may now be calculated.

PROPOSITION 12.2
Let f be a density such that $(x - D)^r f \in L^2(\mathbb{R})$, for $r \geq 1$; then the mean square error of the estimator given by (12.10) satisfies

$$E[\tilde{f}_m(x) - f(x)]^2 = O\left(N^{-1+\frac{6}{6r+1}}\right), \quad m = N^{\frac{6}{6r+1}}$$

uniformly for $x \in \mathbb{R}$.

PROOF The variance of the estimator is given by

$$\nu_m(x) = \sum_{n=0}^{m} E(\tilde{a}_n - a_n)^2 h_n^2(x) \leq \frac{C^4}{N} \sum_{n=0}^{m} (n+1)^{-\frac{1}{3}}$$

$$\leq C_1 \frac{m^{2/3}}{N}$$

by (12.11). The bias estimate uses the fact that $(x + D)h_n = \sqrt{2n}\, h_{n-1}$ (see Chapter 6), which implies that for f satisfying the hypothesis,

$$a_n \sqrt{2n} \sqrt{2n+2} \cdots \sqrt{2n+2r-2}$$

$$= \int_{-\infty}^{\infty} f(x)(x+D)^r h_{n+r}(x)dx$$

$$= \int_{-\infty}^{\infty} (x-D)^r f(x) h_{n+r}(x)dx, \quad n = 0, 1, \cdots .$$

Hence by Bessel's inequality

$$\sum_{n} a_n^2 n^r < \infty$$

and

$$\beta_m^2(x) = \left| \sum_{n=m+1}^{\infty} a_n h_n(x) \right|^2$$

$$\leq \sum_{n=m+1}^{\infty} a_n^2 n^r \sum_{n=m+1}^{\infty} n^{-r} h_n^2(x) = O\left(m^{\frac{5}{6}-r} \right) .$$

We then combine $\nu_m(x)$ and $\beta_m^2(x)$ as in the last section to obtain the conclusion. □

Thus the rate of mean square convergence approaches $O(N^{-1})$ as $r \to \infty$. However, just as with the Fourier series estimator, it is possible to find continuous functions for which the bias does not tend to 0. The hypothesis is satisfied for gamma densities for certain values of r.

12.3 The histogram as a wavelet estimator

In this section we consider an estimator based on the Haar wavelets (Prototype I of Chapter 1). We shall see that this estimator is just the traditional histogram.

The estimator in (12.2) has an alternate expression

$$\tilde{f}_m(x) = \sum_{n=-\infty}^{\infty} \tilde{a}_{mn} 2^{m/2} \phi(2^m x - n)$$

$$\tilde{a}_{mn} = \frac{1}{N} \sum_{i=1}^{N} 2^{m/2} \phi(2^m X_i - n).$$

(12.12)

In this form it can be truncated to a finite sum

$$\tilde{f}_{mk}(x) = \sum_{n=-k}^{k} \tilde{a}_{mn} 2^{m/2} \phi(2^m x - n),$$

(12.13)

which has many of the same convergence properties.

The reproducing kernels $\{q_m(x,y)\}$ constitute a quasi-positive delta sequence that satisfies

$$q_m^2(x,y) = 2^{2m} q^2(2^m x, 2^m y) \leq 2^{2m}.$$

This is easy to show in this case since $\phi(x)$ and $q_m(x,y)$ are just characteristic functions of intervals (indicator functions in statistics terminology).

The estimator $\tilde{f}_m$, as opposed to f^*, is not an unbiased estimator of f. However since the bias is

$$\beta_m(x) = E \tilde{f}_m(x) - f(x)$$

$$= \int q_m(y,x) f(y) dy - f(x),$$

(12.14)

and this converges to 0 for continuous $f(x)$ by Proposition 1.5, it is *asymptotically unbiased*.

The variance of $\tilde{f}_m$ is defined by

$$\nu_m(x) := E \left[\tilde{f}_m(x) - E \tilde{f}_m(x) \right]^2,$$

which again is easy to show, satisfies

$$\nu_m(x) \leq \frac{1}{N} \int q_m^2(y,x) f(y) dy \leq \frac{2^{2m}}{N}.$$

The two expressions may be combined into the *mean square error*, as in the last two sections,

$$MSE = E[\tilde{f}_m(x) - f(x)]^2 = \nu_m(x) + \beta_m^2(x),$$

which if $m = O(\log N)$ converges to 0 uniformly on bounded sets for continuous $f(x)$.

Another useful measure of error is the *integrated mean square error*,

$$IMSE = \int_{-\infty}^{\infty} E[\tilde{f}_m(x) - f(x)]^2 dx, \qquad (12.15)$$

which also converges to 0 if $m = O(\log N)$. The integral of ν_m satisfies

$$\int_{-\infty}^{\infty} \nu_m(x) dx \leq \frac{1}{N} \int_{-\infty}^{\infty} \int_{-\infty}^{\infty} q_m^2(y, x) f(y) dy dx$$

$$= \frac{1}{N} \int_{-\infty}^{\infty} q_m(y, y) f(y) dy \leq \frac{2^m}{N}$$

while that of β_m^2 is

$$\int_{-\infty}^{\infty} \beta_m^2(x) dx = \|f_m - f\|^2,$$

where $f_m = E\tilde{f}_m = \int q_m(y, \cdot) f(y) dy$. Since the MRA $\{V_m\}$ is dense in $L^2(\mathbb{R})$, and f_m is the projection of f on the space V_m, we see that this latter integral converges to 0 for all $f \in L^2(\mathbb{R})$.

Both of these properties are expected of any reasonable density estimator. Another advantage of the wavelet approach is in the calculation of the coefficients in (12.12), which may be expressed as

$$\tilde{a}_{mn} = \frac{2^{m/2}}{N} \Big/ \#\{X_i \epsilon[n2^{-m}, (n+1)2^{-m})\}. \qquad (12.16)$$

We choose $m = \overline{m}$ initially at the finest scale of interest, which would certainly be no finer than that required to put all X_i in different intervals.

We then consider successively coarser choices of scale until we get the coarsest scale of interest, say $m = \underline{m}$ when all X_i belong to the same interval. However we do not use (12.16) to calculate the new $\tilde{a}_{mn}$ but rather use the decomposition algorithm of Mallat. This involves the projection of V_m onto V_{m-1} and W_{m-1} and is given by

$$a_{m-1,k} = 2^{-1/2}(a_{m,2k} + a_{m,2k+1})$$
$$b_{m-1,k} = 2^{-1/2}(a_{m,2k} - a_{m,2k+1}). \qquad (12.17)$$

We may iterate it as often as we want but in practice would stop after we get to $\underline{m}$. In fact we can stop long before it is reached, since the b_{mk}, heretofore unused, give us a measure of the error at each stage. Here's how it works: the finest scale $\overline{m}$ gives us an integrated mean square error

$$e_{\overline{m}} = \int E[\tilde{f}_{\overline{m}}(x) - f(x)]^2 dx,$$

which we cannot improve practically by increasing m. Such an increase would increase the variance without a corresponding reduction in bias.

This may be expressed as

$$
\begin{aligned}
e_m &= \int \mathrm{Var}\, \tilde{f}_m(x)dx + \int \mathrm{Bias}^2 \tilde{f}_m(x)dx \\
&= \int E\left[\frac{1}{N}\sum_{i=1}^{N} q_m(x, X_i)\right]^2 - \frac{1}{N}\int f_m^2(x)dx + \int (f(x) - f_m(x))^2 dx \\
&= \frac{1}{N}\left[\int q_m(x, x)f(x)dx - \sum_k a_{m,k}^2\right] + \sum_{j=m}^{\infty}\sum_k b_{jk}^2 \\
&= \frac{1}{N}\left[\int 2^m \phi(2^m x - [2^m x])f(x)dx - \sum_k a_{m-1,k}^2 - \sum_k b_{m-1,k}^2\right] \\
&\quad + \sum_{j=m}^{\infty}\sum_k b_{jk}^2\,,
\end{aligned}
$$

where we have used the orthogonality to express

$$
\sum_k a_{m,k}^2 = \sum_k a_{m-1,k}^2 + \sum_k b_{m-1,k}^2.
$$

Since $\phi(x - [x]) = 1$, the integral is just 2^m, and the change in error in going from one scale to the next coarser scale is

$$
\begin{aligned}
e_m - e_{m-1} &= \frac{1}{N}\left[2^m - \sum_k a_{m-1,k}^2 - \sum_k b_{m-1,k}^2 - 2^{m-1}\right. \\
&\quad \left. + \sum_k a_{m-1,k}^2\right] - \sum_k b_{m-1,k}^2 \\
&= \frac{2^{m-1}}{N} - \frac{N+1}{N}\sum_k b_{m-1,k}^2\,.
\end{aligned}
$$

That is, the change in error depends only on the wavelet coefficients $b_{n,k}$. We proceed by starting with $\overline{m}$ and checking how much the error changes in going from $\overline{m}$ to $\overline{m} - 1$. Then we iterate this as long as the error is small until we find an m at which the error increases by a larger amount. This will then be our coarsest scale.

In practice we cannot know these coefficients exactly but must use the estimate given by (12.12) for a_{mn}. This in turn will give us the estimated $\tilde{b}_{m-1,n}$ and $\tilde{e}_m$.

Example. A random sample of size 25 based on a $N(0, 1)$ distribution was generated. The finest scale was chosen to be $\overline{m} = 3$, and the values of the estimated coefficients $\tilde{a}_{3,k}$ were calculated. These were used to calculate the wavelet coefficients $\tilde{b}_{m,k}$ and the other scaling coefficients $\tilde{a}_{m,k}$ from (12.17). The results are summarized in Table 1.

Table 1. Total estimated error for different scales m.

m	3	2	1	0	-1	-2	-3
$\sum_k \tilde{b}_{mk}^2$	-	0.208	0.029	0.059	0.026	0.108	0.65
$\sum_k \tilde{a}_{mk}^2$	0.550	0.352	0.323	0.264	0.238	0.130	0.65
$\tilde{e}_m - \tilde{e}_3$	0	0.056	0.006	0.027	0.034	0.136	0.807

In this case the total error increases most rapidly in going from $m = -1$ to $m = -2$, and hence $m = -1$ is probably the appropriate value. This is just a histogram with nodes at $\cdots, -2, 0, 2, 4, \cdots$.

12.4 Smooth wavelet estimators of density

We now return to the general case and assume that $\phi \in S_r$ and satisfies property Z_λ of Chapter 8. We also assume that $f(x)$, the density in question, has a degree of smoothness by requiring it to belong to the Sobolev space H^α for some $\alpha > \frac{1}{2}$.

The estimator will again be given by (12.2) except that $q_m(x, y)$ is the general reproducing kernel of V_m. The other form of the estimator (12.12) and the truncated version (12.13) are equally valid in the general case.

THEOREM 12.3
Let the scaling function $\phi \in S_r$ and satisfy property Z_λ for some $\lambda \geq 1$; let $X_1, X_1, \cdots, X_N$ be an i.i.d. sample from the continuous bounded density $f(x)$; let

$$\tilde{f}_m(x) = \frac{1}{N} \sum_{i=1}^N q_m(x, X_i), \qquad x \in \mathbb{R}.$$

Then

(i) $E|\tilde{f}_m(x) - f(x)|^2 \to 0$, *uniformly on compact sets as $m \to \infty$ and $m = O(\log N)$,*

(ii) *if $f \in H^\alpha$, $\alpha > \lambda + \frac{1}{2}$, $m \approx \frac{\log 2}{2\lambda+1} \log N$,*

(a) $E|\tilde{f}_m(x) - f(x)|^2 = O(N^{-2\lambda/2\lambda+1})$

(b) $\int E|\tilde{f}_m(x) - f(x)|^2 dx = O(N^{-2\lambda/2\lambda+1})$.

PROOF   The variance of the estimator is given by

$$E|\tilde{f}_m(x) - f_m(x)|^2 \leq \frac{1}{N} \int_{-\infty}^{\infty} q_m^2(x,y) f(y) dy$$
$$\leq \frac{1}{N} \|f\|_\infty q_m(x,x) = \frac{1}{N} \|f\|_\infty 2^m q(2^m x, 2^m x)$$
$$= O(2^m/N) \tag{12.18}$$

since $q(x,x)$ is continuous and periodic. The bias converges to 0 uniformly on compact sets since $q_m(x,y)$ is a quasi-positive delta sequence. Hence if $m = O(\log N)$ and $m \to \infty$ then both the bias and the variance converge to 0.

The rate of convergence in (ii) is obtained by using Lemma 8.7 and (12.18). If $f \in H^\alpha$, then the bias is given by

$$|E\tilde{f}_m(x) - f(x)| = \left| \int q_m(x,y) f(y) dy - f(x) \right|$$
$$\leq \|q_m(x,\cdot) - \delta(x,\cdot)\|_{-\alpha} \|f\|_\alpha = O(2^{-m\lambda})$$

by Schwarz's inequality for Sobolev spaces. Hence if $2^m/N \equiv 2^{-2m\lambda}$ then the mean square error is bounded by $O(N^{-2\lambda/2\lambda+1})$.

The same inequalities, though not the same proof, hold for the integrated mean square error.

REMARK 12.1   In some wavelet cases, such as the Meyer wavelets, the parameter λ can be taken arbitrarily large and hence the error rate approaches $O(N^{-1})$.

REMARK 12.2   The estimator may also be expressed in truncated form

$$\tilde{f}_{m,k}(x) = \sum_{n=-k}^{k} \left\{ \frac{1}{N} \sum \phi(2^m X_i - n) \right\} 2^m \phi(2^m x - n),$$

for which the same rate is obtained—provided $k = k(m)$ grows sufficiently fast.

REMARK 12.3   The estimator $\tilde{f}_m$ may also be given in terms of the empiric distribution $f^*(x)$ given by (12.1) as

$$\tilde{f}_m(x) = \int_{-\infty}^{\infty} q_m(x,y) f^*(y) dy = - \int_{-\infty}^{\infty} \frac{\partial q_m(x,y)}{\partial y} F^*(y) dy.$$

COROLLARY 12.4
Let $f(x)$, $\phi(x)$ be as in Theorem 12.3 with $r \geq 1$. Then $\tilde{f}_m(x)$ is almost surely consistent as $m \to \infty$, $m = O(\log N)$ and

$$\tilde{f}_m(x) - f(x) = O\left((N^{-\lambda/(\lambda+1)} \log \log N)^{\frac{1}{2}}\right) \qquad a.s.,$$

for $m = O(\log N/2(\lambda + 1) \log 2)$.

The proof follows the law of the iterated logarithm [Ki], together with the bound

$$|\tilde{f}_m(x) - f(x)| \leq |\tilde{f}_m(x) - f_m(x)| + |f_m(x) - f(x)|$$

$$\leq \sup_y |F^*(y) - F(y)| \int_{-\infty}^{\infty} \left|\frac{\partial q_m}{\partial y}(x, y)\right| dy + O(2^{-m\lambda})$$

$$= O\left(N^{-\frac{1}{2}} |\log \log N|^{\frac{1}{2}} 2^m\right) + O(2^{-m\lambda}).$$

The results given above are asymptotic and not of much use in choosing an appropriate value of the parameter m. Fortunately the techniques of the prototype Haar system carry over to the general case.

The coefficients of our density estimator

$$\tilde{a}_{mk} = \frac{1}{N} \sum_{i=1}^{N} \phi_m(X_i - k2^{-m})$$

and

$$\tilde{b}_{mk} = \frac{1}{N} \sum_{i=1}^{N} \psi_m(X_i - k2^{-m})$$

are formally the same as in the Haar case. The dilation equations give us our decomposition algorithm,

$$a_{m-1,k} = 2^{-\frac{1}{2}} \sum_j c_j a_{m,2k+j}$$

$$b_{m-1,k} = 2^{-\frac{1}{2}} \sum_j c_{1-j}(-1)^j a_{m,2k+j}. \qquad (12.19)$$

Hence we can again proceed as before. We choose a finest scale $\overline{m}$ of interest; calculate $\tilde{a}_{m,k}$ from the data and then use (12.19) to find $\tilde{a}_{m-1,k}$ and $\tilde{b}_{m-1,k}$. We keep decreasing m until the integrated mean square error takes a jump.

However there is one difficulty. In the case of the prototype, $\tilde{a}_{m,k}$ was obtained by merely counting the number of sampled values. In general $\phi(x)$ is not a constant function, and mere counting will not work. In fact $\phi(x)$ can be a very wild function and is usually not even given by a formula. Rather the properties of ϕ must be obtained from the dilation equation. The values of $\phi(x)$ may be found by beginning with $\phi^{(0)}(x)$ the prototype and then approximating $\phi(x)$ by $\phi^{(j)}(x)$ where

$$\phi^{(j+1)}(x) = \sqrt{2} \sum_k c_k \phi^{(j)}(2x - k), \qquad j = 0, 1, 2, \cdots. \qquad (12.20)$$

This scheme works for certain choices of the c_k's. (see [D] or [V]), that is

$$\phi^{(j)}(x) \rightarrow \phi(x) \quad \text{in} \quad L^2(\mathbb{R}),$$

and $\{\phi(x - n)\}$ is orthonormal.

We may use (12.20) directly to define the coefficients $\tilde{a}_{mk}$. Indeed we have

$$a_{m,n}^{(j+1)} = 2^{-\frac{1}{2}} \sum_k c_k a_{m-1,2n+k}^{(j)} \qquad j = 0, 1, \cdots. \qquad (12.21)$$

Thus we begin with $\tilde{a}_{m,k}^{(0)}$, obtained by counting, and use the recursive definitions (12.20) and (12.21) to approximate $\tilde{f}_m(x)$ by

$$\tilde{f}_m^{(j)}(x) = \sum_n \tilde{a}_{m,n}^{(j)} \phi^{(j)}(x - n) \qquad j = 0, 1, \cdots. \qquad (12.22)$$

Two ways of choosing m are possible: one is to begin with the m^* obtained by using the prototype histogram. The appropriate choice for (12.22) would be

$$m^{**} = m^* + j.$$

This gives us a finer scale than with the prototype.

The other approach emulates that of section 3. Choose $\overline{m}$ and calculate $\tilde{a}_{\overline{m},k}$. Then use (12.20) until the integrated mean square error takes a jump. The calculation of e_m is very similar. Indeed the first expression for e_m is still valid and

$$\begin{aligned} e_m - e_{m-1} = &\frac{1}{N} \int_{-\infty}^{\infty} 2^m q(2^m x, 2^m x) f(x) dx \\ &- \frac{N+1}{N} \sum_k b_{m-1,k}^2. \end{aligned} \qquad (12.23)$$

By (12.19) we again have

$$\sum_k a_{m,k}^2 = \sum_k a_{m-1,k}^2 + \sum_k b_{m-1,k}^2,$$

which enables us to find the best m by decreasing it until e_m takes a jump. The integral in (12.23) may be approximated by $\frac{1}{N}2^m \sum_{i=1}^{N} q(2^m X_i, 2^m X_i)$ resulting from replacing $f(x)$ by the empiric distribution $f^*(x)$ of (12.1).

12.5 Local convergence

If the distribution of the sample does not have a density, and we form the density estimator (12.2), it cannot of course be integrated mean square consistent. However it will be locally mean square consistent at points at which the distribution function $F(x)$ is differentiable. This is a consequence of Lemma 8.4, which we repeat.

LEMMA 8.4
Let $q_m(x,t)$ be the RK of V_m, $\phi \in \phi_r$, $r \geq 1$, $0 \leq \alpha \leq r$; then

$$K_m(x,t) = \frac{(x-t)^\alpha}{\alpha!} \frac{\partial^\alpha}{\partial t^\alpha} q_m(x,t)$$

is a quasi-positive delta sequence.

As an immediate corollary we have

COROLLARY 12.5
Let the distribution function $F(x)$ have a derivative at x_0. Then

$$E(\tilde{f}_m(x_0)) \to F'(x_0)$$

as $m \to \infty$.

The proof is similar to the previous ones since

$$E\tilde{f}_m(x_0) = \int q_m(x_0,t)dF(t)$$
$$= \int (x_0 - t)\frac{\partial q_m(x_0,t)}{\partial t} \frac{F(t) - F(x_0)}{t - x_0}dt.$$

This must converge to

$$\lim_{t \to x_0} \frac{F(t) - F(x_0)}{t - x_0} = F'(x_0).$$

Thus $\tilde{f}_m(x_0)$ is asymptotically unbiased and, since the variance has the same bound as in the last section, is mean square consistent as well for $m = O(\log N)$.

12.6 Positive density estimators

Since only the trivial case of the Haar system gives scaling functions
that are positive [J1], the estimators of the previous sections have not
been density functions themselves. Another estimator based on wavelets
is possible however. It uses the scaling function basis $\{\phi_{mn}\}$ of V_m to
approximate the characteristic function $\theta(w)$, i.e., the Fourier transform,
rather than the density function. If the density function $f(x)$ is bounded,
then $\theta(w) \in L^2(\mathbb{R})$ and its projection on V_m may be given by

$$\theta_m(w) = \int_{-\infty}^{\infty} q_m(w,\zeta)\theta(\zeta)d\zeta$$

$$= \sum_n a_{mn} 2^{m/2}\phi(2^m w - n).$$

The Fourier transform of θ_m is

$$\hat{\theta}_m(x) = \sum_n a_{mn} 2^{-m/2}\hat{\phi}(2^{-m}x)e^{-ixn2^{-m}},$$

which may be used to obtain an approximation to $f(x)$

$$f_m(x) = \frac{1}{2\pi}\hat{\theta}_m(-x)$$

$$= \frac{1}{2\pi}\hat{\phi}(-2^{-m}x)2^{-m/2}\sum_n a_{mn}e^{inx2^{-m}}. \tag{12.24}$$

For the scaling function ϕ given by

$$\hat{\phi}^2(x) = \int_{x-\pi}^{x+\pi} h,$$

for symmetric, positive function h supported on $[-\epsilon, \epsilon]$ (Example 7 of
Chapter 3) we have $\phi(x) > 0$ for $|x| < \pi + \epsilon$ and $\phi(x) = 0$ for $|x| > \pi + \epsilon$
and $\phi(x) = \phi(-x)$. Hence in this case $f_m(x)/\hat{\phi}(-2^{-m}x)$ is a function in
$L^2(-2^{m+1}\pi, {}^{m+1}\pi)$. Furthermore, $f_m(x) \geq 0$; to see this we express θ_m
as

$$\theta_m(w) = \frac{1}{2\pi}\int \hat{q}_m(w,y)\overline{\hat{\theta}(y)}dy$$

by using the isometry property of the Fourier transform. Then f_m becomes, for $m = 0$,

$$f_0(x) = \frac{1}{2\pi} \int_{-\pi-\epsilon}^{\pi+\epsilon} \sum_n \hat{\phi}(x)\hat{\phi}(y)e^{-in(x+y)}f(-y)dy$$

$$= \int_{-\pi-\epsilon}^{\pi+\epsilon} \hat{\phi}(x)\hat{\phi}(y)\delta^*(x+y)f(-y)dy$$

$$= \hat{\phi}(x)\sum_k \hat{\phi}(x+2\pi k)f(x+2\pi k),$$

each of the terms of which is nonnegative. (Here δ^* is the periodic extension of the delta function.)

We now estimate $f_0(x)$ – and by a change of scale, $f_m(x)$ – by using the Fejer kernel F_n on $[-\pi, \pi]$ to estimate the periodic part. This leads to the estimator

$$\tilde{f}_{0,n}(x) = \hat{\phi}(x)\frac{1}{N}\sum_i F_n(x - X_i)\hat{\phi}(X_i). \tag{12.25}$$

For $|x| < \pi - \epsilon$, we have

$$E\tilde{f}_{0,n}(x) = \hat{\phi}(x)\int_{-\infty}^{\infty} F_n(x - t)f(t)\hat{\phi}(t)dt$$

$$= \hat{\phi}(x)\sum_x \int_{-\pi}^{\pi} F_n(x - t)f(t + 2\pi k)\hat{\phi}(t + 2\pi k)dt \rightarrow f_0(x)$$

at points of continuity as $n \rightarrow \infty$. Since $f_0(x) = f(x)$ for $|x| < \pi - \epsilon$, we have asymptotic unbiasedness. For other values of x we use a change of scale and the resulting estimator

$$\tilde{f}_{m,n}(x) = 2^{-m}\tilde{f}_0(2^{-m}x). \tag{12.26}$$

Thus we have a nonnegative estimator whose Fourier transform belongs to V_m. The rate of convergence will be relatively slow since even for C^∞ functions $(F_n * f)(x) - f(x) = O(n^{-1})$.

12.7 Other estimation with wavelets

Other types of nonparametric estimation in statistics involve (i) spectral density estimation, (ii) nonparametric regression, and (iii) mixture problems. Spectral density estimation involves techniques similar to those for probability density, except that periodic estimators must be used. Nonparametric regression can be based on probability density estimation but requires multivariate estimation. The same is true for estimation of mixtures except that a sample, instead of being taken from the density, is taken from a mixture of the unknown density with another distribution.

12.7.1 Mixture problems

In many experimental settings, it is impossible to obtain measurements of the desired random variable. Rather the only measurements obtainable are corrupted by noise. If this noise is additive and independent we have a mixture problem. The measurements are those of the form

$$Y = X + Z$$

where X is the desired value, and Z is a random variable with a known distribution that we assume to have a density $k(x)$. Then the density of Y is given by

$$f(y) = \int k(y - x)g(x)dx = (k * g)(y).$$

Given a sample $Y_1, Y_2, \cdots, Y_N$, we wish to estimate $g(x)$. This estimation is difficult since it leads to an ill-posed problem. Even if we succeed in estimating $f(y)$ well, this is no guarantee that we have an estimator of $g(x)$. This is true even when the problem is identifiable, i.e., when the integral operator is one to one.

Many solutions have been proposed, but none are completely satisfactory. The most widely used method seems to be the E-M algorithm [D-L-R]. This however leads to estimators of the form

$$\tilde{g}(x) = \sum a_i \delta(x - x_i),$$

which must be smoothed to obtain density estimations. For natural exponential families with quadratic variance functions, a method based on orthogonal polynomials may be used [W-H]. These polynomials are mapped into another biorthogonal system under the integral operator with kernel $k(y \mid x)$ (in our case, a difference kernel). This is the procedure we imitate with wavelets.

As usual we first approximate $g(x)$ by an element $g_m(x) \in V_m$, and we denote by U_m the range of the operator K with kernel $k(y - x)$ when the domain is V_m, i.e.,

$$U_m = KV_m.$$

We may assume, by changing the scale if necessary, that $m = 0$. Then if $g_0 \in V_0$ is given by

$$g_0(x) = \sum_n a_n \phi(x - n)$$

we have $f_0 \in U_0$ given by

$$f_0(y) = \sum a_n \int k(y-x)\phi(x-n)dx$$
$$= \sum a_n \int k(y-n-x)\phi(x)dx$$
$$= \sum a_n \alpha(y-n)$$

where $\alpha = k * \phi$. This always holds since K is a continuous operator on $L^2(\mathbb{R})$.

We now take Fourier transforms and observe that

$$\hat{f}_0(w) = \hat{k}(w)\hat{g}_0(w).$$

Hence if the problem is to be identifiable $\hat{k}(w) \neq 0$ a.e., but we shall assume further that $\hat{k}(w) \neq 0$ everywhere on $\mathbb{R}$.

We denote by $\beta(x)$ the function given by its Fourier transform as

$$\hat{\beta}(w) = \hat{\phi}(w)/\overline{\hat{k}(w)}. \tag{12.27}$$

For a Meyer type scaling function (Example 7 of Chapter 3) $\hat{\phi}(w)$ has support in $\left[-\frac{4\pi}{3}, \frac{4\pi}{3}\right]$ and hence $\hat{\beta} \in L^2(\mathbb{R})$. Since $\hat{\beta}$ can also be written as

$$\hat{\beta}(w) = \hat{k}(w)\left\{\frac{\hat{\phi}(w)}{|\hat{k}(w)|^2}\right\},$$

$\beta(x)$ also is in the range of the operator K.

LEMMA 12.6
Let $\alpha = k * \phi$, and let β be given by (12.27). Then $\{\alpha(y-n), \beta(y-m)\}$ is a biorthogonal sequence, and $\{\alpha(y-n)\}$ is a Riesz basis of U_0.

PROOF   The biorthogonality follows from the orthonormality of $\{\phi(y-n)\}$; we have

$$\int \alpha(y-n)\overline{\beta(y-m)}dy = \frac{1}{2\pi}\int \hat{\alpha}(w)\overline{\hat{\beta}(w)}e^{-inw+imw}dw$$
$$= \frac{1}{2\pi}\int \hat{k}(w)\hat{\phi}(w)\frac{\overline{\hat{\phi}(w)}}{\hat{k}(w)}e^{-i(n-m)w}dw$$
$$= \int \phi(y-n)\phi(y-m)dy.$$

To show $\{\alpha(t-n)\}$ is a Riesz basis we need only show that there are constants $C' \geq C > 0$ such that

$$C\sum|a_k|^2 \leq \left\|\sum a_k\alpha(t-k)\right\|^2 \leq C'\sum|a_k|^2$$

for each $\{a_k\} \in \ell^2$ [M]. By the Fourier transform we have

$$\left\|\sum a_k \alpha(t-k)\right\|^2 = \frac{1}{2\pi} \int |\hat{\alpha}(w)|^2 \left|\sum_k a_k e^{-ikw}\right|^2 dw$$

$$= \frac{1}{2\pi} \int_0^{2\pi} \sum_k |\hat{\alpha}(w+2\pi k)|^2 \left|\sum_k a_k e^{-ikw}\right|^2 dw$$

$$= \frac{1}{2\pi} \int_0^{2\pi} \sum_k \left|\hat{\phi}(w+2\pi k)\right|^2 \left|\hat{k}(w+2\pi k)\right|^2 \left|\sum_k a_k e^{-ikw}\right|^2 dw$$

$$\leq \frac{1}{2\pi} \int_0^{2\pi} \sum_k \left|\hat{\phi}(w+2\pi k)\right|^2 \left|\sum_k a_k e^{-ikw}\right|^2 dw$$

$$= \frac{1}{2\pi} \int_0^{2\pi} \left|\sum_k a_k e^{-ikw}\right|^2 dw = \sum_k |a_k|^2 \qquad (12.28)$$

since $|\hat{k}(w)| \leq 1$ and $\sum \left|\hat{\phi}(w+2\pi k)\right|^2 = 1$. Moreover, the series

$$\sum_k \left|\hat{\phi}(w+2\pi k)\right|^2 \left|\hat{k}(w+2\pi k)\right|^2$$

has only two nonzero terms. Hence (12.28) may be expressed as

$$\left\|\sum_k a_k \alpha(t-k)\right\|^2 = \frac{1}{2\pi} \int_{4\pi/3}^{4\pi/3} \left|\hat{\phi}(w)\right|^2 \left|\hat{k}(w)\right|^2 \left|\sum_k a_k e^{-ikw}\right|^2 dw$$

$$\geq \inf_{|w|\leq 4\pi/3} |\hat{k}(w)|^2 \frac{1}{2\pi} \int_{-\infty}^{\infty} |\hat{\phi}(w)|^2 \left|\sum_k a_k e^{ikw}\right|^2 dw$$

$$= C \left\|\sum a_k \phi(t-k)\right\|^2 = C \sum |a_k|^2$$

where

$$C = \inf_{|w|\leq 4\pi/3} |\hat{k}(w)|^2. \qquad \square$$

We now return to the case of general m and redefine our bases. We replace $\phi_0(w)$ by $\phi_m(w) = \phi(2^m w)2^{m/2}$ and α_0, β_0 by

$$\alpha_m = k * \phi_m,$$
$$\hat{\beta}_m(w) = \hat{\phi}_m(w) / \hat{k(w)}. \qquad (12.29)$$

The spaces $U_m = K V_m$ have the same nested property that the V_m's do,

$$\cdots \subset U_m \subset U_{m+1} \subset \cdots \subset L^2(\mathbb{R})$$

and

$$\bigcap_n U_m = \{0\}.$$

Furthermore

$$\overline{\bigcup_m U_m} = L^2(\mathbb{R})$$

since any $f \in L^2(\mathbb{R})$ whose Fourier transform $\hat{f}(w)$ has compact support belongs to some U_m. This follows since for such f

$$\hat{f}(w)/\hat{k}(w)\hat{\phi}_m(w) \in L^2\left[-2^{m+1}\pi, 2^{m+1}\pi\right]$$

for m sufficiently large.

Notice however that while $\{\alpha_m(x - n2^{-m})\}$ is a Riesz basis of U_m it is not obtained from α_{m+1} by a dilation equation.

This gives us the tool we need to construct our estimator. Given the sample $Y_1, \ldots, Y_N$, we define

$$\tilde{a}_{km} := \frac{1}{N} \sum_{i=1}^N \beta_m(Y_i - k2^{-m})$$

$$\tilde{f}_m(x) := \sum_k \tilde{a}_{km}\alpha_m(x - k2^{-m}) \tag{12.30}$$

$$\tilde{g}_m(x) := \sum_k \tilde{a}_{km}\phi_m(x - k2^{-m}).$$

Then we find for $m = 0$

$$
\begin{aligned}
E\tilde{g}_0(x) &= \sum_k \phi(x - k) \int \beta(t - k)f(t)\,dt \\
&= \sum_k \phi(x - k)\frac{1}{2\pi} \int \hat{\beta}(w)\overline{\hat{f}(w)}\,e^{-iwk}\,dw \\
&= \sum_k \phi(x - k)\frac{1}{2\pi} \int \left(\hat{\phi}(w)/\hat{k}(w)\right)\overline{\hat{k}(w)}\,\overline{\hat{g}(w)}\,e^{-iwk}\,dw \\
&= \sum_k \phi(x - k) \int \phi(t - k)g(t)\,dt = g_0(x).
\end{aligned}
\tag{12.31}
$$

By a change of scale we get $E\tilde{g}_m(x) = g_m(x)$. We also have another expression for $\tilde{g}_m$,

$$\tilde{g}_m(x) = \frac{1}{N} \sum_{i=1}^N \sum_k 2^{m/2}\beta_m(X_i - 2^{-m}k)\phi(2^m x - k).$$

Hence the bias $g_m(x) - g(x)$ is exactly the same as it is for density estimators, i.e., $O(2^{-m\lambda})$ for $g \in H^\alpha$, the Sobolev space.

However, the variance is another matter. It satisfies

$$E\left[\tilde{g}_m(x) - g_m(x)\right]^2 = \text{Var}_m$$

$$\leq \frac{1}{N} \int_{-\infty}^{\infty} \left\{ \sum_k \beta_m(t - 2^{-m}k)\phi_m(x - 2^{-m}k) \right\}^2 f(t)dt,$$

and

$$\beta_m(t) = \frac{1}{2\pi} \int_{-2^m \frac{4\pi}{3}}^{2^m \frac{4\pi}{3}} \frac{\hat{\phi}_m(w)}{\hat{k}(w)} e^{iwt} dw$$

$$= \frac{2^{m/2}}{2\pi} \int_{-4\pi/3}^{4\pi/3} \frac{\hat{\phi}(w)e^{i2^m wt}}{\hat{k}(2^m w)} dw.$$

Hence we have

$$|\beta_m(t)| \leq 2^{m/2}\frac{4}{3}C_m, \quad C_m = \frac{1}{\inf\limits_{|w| \leq 4\pi/3} |k(2^m w)|},$$

and the variance satisfies

$$\text{Var}_m \leq \frac{1}{N}(2^{m/2}(4/3)C_m)^2 \left\{ \sum_k 2^{m/2}|\phi(2^m x - k)| \right\}^2.$$

Since $|\phi(x)| \leq \frac{C}{(1+|x|)^2}$, we find

$$\text{Var}_m \leq \frac{1}{N}2^{2m}C_m'(1 + |x|)^{-2}.$$

We have proved

PROPOSITION 12.7
Let $Y_1, Y_2, \cdots Y_n$ be a random sample from a mixture density $f = k * g$, where $g \in H^\alpha$, $\alpha > \lambda + \frac{1}{2}$; let $\alpha_m(x), \beta_m(x)$ be given by (12.29). Then $\tilde{g}_m(x)$ (12.30) satisfies
   (i) $|E\tilde{g}_m(x) - g(x)| = O(2^{-m\lambda})$ *as* $m \to \infty$,
   (ii) *there is $\{m(N)\}$ such that $E|\tilde{g}_m(x) - g(x)|^2 \to 0$, $N \to \infty$, $m = m(N)$.*

12.7.2 Spectral density estimation

We start with a discrete stationary random process $\{X(t), t \in \mathbb{Z}\}$ with zero mean and finite covariance. We wish to estimate the spectral density

$$f(\lambda) = \frac{1}{2\pi} \sum_{k=-\infty}^{\infty} C_k e^{-ik\lambda}$$

where $C_k = E(X(t)X(t+k))$ is the covariance. We assume it is observed at $t = 1, 2, \cdots, N$ and form the periodogram

$$I_N(\lambda) = \frac{1}{2\pi} \sum_{k=-N+1}^{N-1} \tilde{C}_k e^{-i\lambda k}$$

where

$$\tilde{C}_k = \begin{cases} \frac{1}{N} \sum_{t=1}^{N-|k|} X(t+|k|)X(t), & |k| < N \\ 0, & |k| \geq N. \end{cases}$$

While $I_N(\lambda)$ is an asymptotically unbiased estimator of $f(\lambda)$ since

$$E(I_N(\lambda)) \equiv \frac{1}{2\pi} \sum_{k=-N+1}^{N-1} C_k e^{-i\lambda k},$$

its variance gives problems, and it must be smoothed by a spectral window $W_M(\lambda)$ usually consisting of a trigonometric polynomial. The estimator then is

$$\tilde{f}_{MN}(\lambda) = (W_M * I_N)(\lambda).$$

An estimator of this kind may be based on periodic wavelets since the spectral density is a periodic function. These are constructed from the usual wavelets by periodization as in Chapter 7. This sequence $\{g_n\}$ of periodic wavelets is composed of periodic functions of period 1 which form an orthonormal basis of $L^2(0, 1)$. The periodized versions of V_m become V_m^p and have a basis consisting of

$$g_0, g_1, \cdots, g_{2^m-1}.$$

The reproducing kernel for V_m^p is

$$q_m^p(x, y) = \sum_{k=0}^{2^m-1} \phi_{m,k}^p(x) \phi_{m,k}^p(y)$$

$$= \sum_{k=0}^{2^m-1} g_k(x) g_k(y).$$

It is again fairly easy to show that these kernels form a quasi-positive delta sequence on $(0,1)$ as $m \to \infty$. The associated spectral density estimator is

$$\tilde{f}_{m,N}(\lambda) = \frac{1}{2\pi} \int_0^{2\pi} q_m^p \left(\frac{\lambda}{2\pi}, \frac{y}{2\pi} \right) I_N(y) dy, \quad m = 0, 1, 2, \cdots . \quad (12.32)$$

This does not exactly correspond to the usual spectral window, which involves convolution, but similar arguments to those used for convolution can be used. In particular, $\frac{1}{2\pi}q_m^p\left(\frac{x}{2\pi}, \frac{y}{2\pi}\right)$ is a delta sequence of class Γ_1 [W13] (with a slight modification of definition). Hence, the estimator (12.32) will be mean square consistent with rate $O(N^{-2/3})$ for smooth spectral densities of Gaussian processes.

There is no advantage of these estimators over the others as far as convergence is concerned, since most spectral windows also lead to quasi-positive delta sequences. There may be an advantage in estimating spectral density functionals of processes with a periodic component. However, this has not as yet been explored.

12.7.3 Regression estimators

Any density estimator on $\mathbb{R}^d$ leads to an estimator of the nonparametric regression function

$$r(x) = E(Y \mid X = x).$$

In particular the wavelet estimators can be used, but we need first to consider the multivariate estimators. For $d = 2$ the estimator is

$$\tilde{f}_m(x, y) = \frac{1}{N}\sum_i q_m(x, X_i)q_m(y, Y_i)$$

where $q_m(x, t)$ is the reproducing kernel of V_m. The estimator of $r(x)$ is

$$\tilde{r}_m(x) = \frac{1}{N}\frac{\sum_{i=1}^N Y_i q_m(x, X_i)}{\frac{1}{N}\sum_{i=1}^N q_m(x, X_i)} = \frac{\tilde{\nu}_m(x)}{\tilde{f}_m^1(x)}. \qquad (12.33)$$

The numerator of (12.33) may be expressed as

$$\begin{aligned}
\tilde{\nu}_m(x) &= \int y\tilde{f}_m(x, y)dy \\
&= \frac{1}{N}\sum_i q_m(x, X_i)\int y q_m(y, Y_i)dy \\
&= \frac{1}{N}\sum_i q_m(x, X_i)Y_i.
\end{aligned}$$

This follows from the property that

$$\int y^k q_m(x, y)dy = x^k, \quad k = 0, 1, 2, \cdots, r$$

where our $\phi(x)$ is r-regular, $r \geq 1$.

Hence the estimator is similar to kernel estimators of the regression function and by the same arguments can be shown to be mean square consistent [P-R].

12.8 Problems

1. In the usual kernel estimator of density, a difference kernel is used in place of the reproducing kernel in (12.2). That is, if $k(x)$ is a density function,

$$\tilde{f}_m(x) = \frac{m}{N} \sum_{i=1}^{N} k(m(x - X_i))$$

is the estimator. If f and k are both continuous with compact support then show that

$$E|\tilde{f}_m(x) - f(x)|^2 \to 0 \quad \text{uniformly if} \quad \frac{m}{N} \to 0$$

by

(i) showing that $mk(m(x - y))$ is a positive delta sequence.

(ii) showing that the bias of $\tilde{f}_m(x)$ converges to $f(x)$ uniformly.

(iii) finding a bound on the variance.

2. Let $f(x) = \frac{1}{4} \cos \frac{x}{2}$, $|x| \leq \pi$, $f(x) = 0$, $|x| > \pi$. Generate a random sample $X_1, X_2 \cdots X_N$ drawn from a distribution with this density by

(i) generating random numbers in $[0,1]$ $Y_1, Y_2, \cdots, Y_N$.

(ii) setting $X_i = F^{-1}(y_i)$, where $F(x)$ is the distribution function, $F(x) = \int_{-\pi}^{x} f(t)dt = \frac{1}{2} + \frac{1}{2} \sin x$, $-\pi \leq x \leq \pi$.

3. Find the density estimator $\tilde{f}_m^{(1)}(x)$ of (12.5) and $\tilde{f}_m^{(2)}$ of (12.6) for a sample size $N = 50$ from Problem 2 for values of $m = 1$, 5, 25.

4. Use the sample of Problem 2 and the Hermite series density estimator of (12.10) to approximate $f(x)$ (using the same values of m).

5. Generate a sample of size 50 from a normal distribution with mean $\mu = 1$ and variance $\sigma = 1$. Use the sample and find the Hermite series estimator for $m = 1, 5$, and 25.

6. Find the histogram estimator with the sample of Problem 5 for $m = 0, 1, 2$ in (12.12).

7. Use the estimator for $m = 2$ in Problem 5 and the decomposition algorithm (12.17) to obtain the estimators for $m = 0$ and $m = 1$.

8. (If you don't have a computer.) Draw 25 numbers, between 0 and 1, from a random number table. Partition the interval $(0,1)$ into 2^3 equal intervals and use (12.16) to determine $\tilde{a}_{3n}$. Find $\tilde{a}_{2,n}$ and $\tilde{a}_{1,n}$ from (12.17), and sketch the resulting histogram.

13

Orthogonal Systems and Stochastic Processes

One of the most elegant uses of orthogonal systems is in conjunction with continuous stochastic processes. Such a process $\{X(t), t \in I\}$ is a family of random variables indexed by the parameter t [Ro, p.91]. For processes that are continuous in mean square, i.e.,

$$E|X(t)|^2 < \infty$$

and

$$E|X(t) - X(s)|^2 \to 0 \quad \text{as} \quad t \to s,$$

there is a theory, the Karhunen-Loève theory, that converts the process into an uncorrelated sequence of random variables. This is done by using a certain orthogonal system, expanding $X(t)$ in terms of it, and showing the coefficients are uncorrelated.

In this chapter we first present this theory. We then consider wavelet expansions of stationary processes and show that for certain wavelets, the coefficients of the expansion have negligible correlation for different scales. We then introduce a modification of the wavelets, which leads to uncorrelated coefficients. Finally we show that for certain nonstationary processes the wavelets may be chosen to give uncorrelated coefficients.

13.1 K-L expansions

In this section we introduce elements of the Karhunen-Loève (K-L) theory. A more complete readable exposition may be found in [Ro, p.203]. We shall assume a background in stochastic processes sufficient for a number of manipulations involved. However they should be understandable, in a formal sense, to persons without such a background.

Let $\{X(t), t \in [0, T]\}$ be a real process that is continuous in mean square and has zero mean $(E(X(t)) = 0)$. Then the covariance function given by

$$r(t, s) = E(X(t)X(s)), \qquad s, t \in [0, T],$$

is a continuous function on $[0, T] \times [0, T]$. If $0 < T < \infty$, then it may be used as the kernel of an integral operator

$$(Rf)(t) := \int_0^T r(t, s)f(s)ds, \qquad (13.1)$$

which will be self adjoint and completely continuous (compact). As such it will have a countable set of eigenvalues $\{\lambda_i\}$ each of which has a finite number of associated eigenfunctions $\phi_i(t)$ that may be assumed to be orthonormal [R-N, p.227]. Then for each g in the range of R (as an operator on $L^2([0, T])$), we have the convergent expansion

$$g(x) = \sum_i \langle g, \phi_i \rangle \phi_i(x), \qquad x \in [0, T]. \qquad (13.2)$$

Moreover this series converges uniformly on $[0, T]$ [R-N, p.244].

We now return to our random process $X(t)$ and expand it in terms of the $\{\phi_i\}$. This involves the random integral

$$X_i = \int_0^T X(t)\phi_i(t)dt, \qquad (13.3)$$

which exists whenever

$$\int_0^T E(|X_t|)dt < \infty, \qquad [\text{Ro, p.205}].$$

But $E|X_t| \leq \{EX_t^2\}^{1/2} = r^{1/2}(t, t)$, which is bounded on $[0, T]$, and hence this integral is finite. The integral (13.3) satisfies the usual properties of integrals.

The X_i in (13.3) are uncorrelated since

$$E(X_i X_j) = E\left\{ \int_0^T X(t)\phi_i(t)dt \int_0^T X(s)\phi_j(s)ds \right\}$$

$$= \int_0^T \int_0^T E(X(t)X(s))\phi_i(t)\phi_j(s)dt\,ds \qquad (13.4)$$

$$= \int_0^T \int_0^T r(t, s)\phi_i(t)\phi_j(s)dt\,ds = \lambda_j \delta_{ij},$$

by a version of Fubini's theorem. We also need to know that $X(t)$ can be represented by such a series with these coefficients. It can be, in the mean square sense, since

$$E\left[X(t) - \sum_{i=0}^{N} X_i \phi_i(t)\right]^2 = r(t,t) - 2\sum_{i=0}^{N} E(X(t)X_i)\phi_i + \sum_{i=0}^{N} \lambda_i \phi_i^2(t)$$

$$= r(t,t) - \sum_{i=0}^{N} \lambda_i \phi_i^2(t). \tag{13.5}$$

By (13.4) the $\lambda_j \geq 0$, and hence we can invoke Mercer's theorem [R-N, p.245], which says that

$$r(t,s) = \sum_i \lambda_i \phi_i(t)\phi_i(s)$$

where the convergence is uniform. Hence the last expression in (13.5) converges to 0 uniformly and

$$X(t) = \sum_i X_i \phi_i(t) \tag{13.6}$$

in the mean square sense.

Example: Brownian Bridge

The Wiener process or "Brownian motion" is a Gaussian process on $[0, \infty)$ such that $W(0) = 0$ and $r(t,s) = \min(t,s)$, $0 < t, s$. The Brownian bridge is a similar process on [0,1] with both $X(0) = 0$ and $X(1) = 0$ with $r(t,s) = \begin{cases} t(1-s), & t < s \\ s(1-t), & s < t \end{cases}$. The eigenvalue problem $R\phi = \lambda\phi$ may, in this case, be converted to a differential equation since

$$\lambda\phi'(t) = \int_0^t -s\phi(s)ds + \int_t^1 (1-s)\phi(s)ds,$$

and

$$\lambda\phi''(t) = -t\phi(t) - (1-t)\phi(t) = -\phi(t).$$

Since $\phi(0) = \phi(1) = 0$, the solution is given by

$$\lambda_n = 1/\pi^2 n^2, \quad \phi_n(t) = \sqrt{2}\sin\pi nt, \quad n = 1, 2, \cdots.$$

Hence we have

$$X(t) = \sum_{n=1}^{\infty} X_n \sqrt{2}\sin\pi nt$$

where
$$X_n = \int_0^1 X(t)\sqrt{2}\, \sin \pi n t \, dt.$$

Then the X_n in this case are not only uncorrelated but, since the process is Gaussian, are in fact independent.

While this K-L theory is elegant, its utility is limited by the difficulty of finding the eigenfunctions. Unless they are also, as in the Brownian bridge example, eigenfunctions of a differential operator, only approximations are obtained. Also in some useful cases in which $X(t)$ is stationary, the interval of definition is infinite. Then $r(s,t)$ is no longer a Hilbert Schmidt kernel (i.e., belongs to $L^2(I \times I)$), and the eigenvalue results do not hold.

13.2   Stationary processes and wavelets

A (<u>wide-sense</u>) <u>stationary process</u> is one for which $E(X(t)) = c$ and

$$E(X(t)X(s)) = E(X(t-s)X(0)),$$

and hence
$$r(t,s) = r(t-s,0) = R(t-s), \quad t,s \in \mathbb{R},$$

depends only on the difference between t and s. Such processes correspond to many physical phenomena. The function $R(t)$ is bounded since

$$|R(t)| = |E(X(t)X(0))| \le \{E(X^2(t))E(X^2(0))\}^{\frac{1}{2}} = R^{\frac{1}{2}}(0)R^{\frac{1}{2}}(0).$$

If $X(t)$ is mean square continuous as well, then $R(t)$ is continuous. The <u>power spectrum</u> of $X(t)$ is the Fourier transform $\hat{R}(\omega)$ of $R(t)$. This may not be an ordinary function but when it is, is called the <u>spectral density</u>.

We will find the wavelet expansion of $X(t)$ with respect to the Meyer type wavelets (Example 7 of Chapter 3). Their scaling function is given by

$$\hat{\phi}(\omega) = \left\{ \int_{\omega-\pi}^{\omega+\pi} dP \right\}^{\frac{1}{2}}, \; P \text{ a probability measure}$$

$$\text{supported on } [-\epsilon, \epsilon], \epsilon \le \frac{\pi}{3}$$

and the mother wavelet by

$$\hat{\psi}(\omega) = e^{-i\omega/2}\left\{\hat{\phi}(\omega + 2\pi) + \hat{\phi}(\omega - 2\pi)\right\}\hat{\phi}\left(\frac{\omega}{2}\right).$$

First however we observe that the approximation of $X(t)$ by $X_m(t)$ where

$$X_m(t) = \sum_n a_{mn}\phi_{mn}(t), \quad a_{mn} = \int X(t)\phi_{mn}(t)dt$$

is mean square consistent for any $\phi \in S_r$. That is,

$$E[X(t) - X_m(t)]^2 \to 0 \quad \text{as} \quad m \to \infty, \, t \in \mathbb{R}.$$

This follows from the fact that $\{q_m(x,t)\}$ is a quasi-positive delta sequence (Chapter 8). Indeed we have

$$E[X(t) - X_m(t)]^2 = E[X^2(t)] - 2E[X(t) - X_m(t)] + E[X_m^2(t)]$$

$$= R(0) - 2\int R(t-s)q_m(s,t)ds$$

$$+ \int\int R(u-s)q_m(s,t)q_m(t,u)du\,ds \quad (13.7)$$

where $q_m(s,t)$ is the reproducing kernel of V_m. (Note however that the sample paths are not in $L^2(\mathbb{R})$ [Ms] and hence $X_m \notin V_m$; nonetheless the interchange of integrals in (13.7) is valid since q_m is rapidly decreasing and $X(t)$ is bounded.)

Since $R(u-s)$ is continuous and in $L^2(\mathbb{R})$, we have

$$\int R(u-s)q_m(s,t)ds \to R(u-t)$$

uniformly on bounded sets and boundedly for all $u \in \mathbb{R}$. Therefore (13.7) converges to 0 as $m \to \infty$.

We can also obtain rates of convergence under additional hypotheses on $X(t)$ by methods similar to those in Chapter 8. We express $X_m(t)$ in the wavelet series as

$$X_m(t) = \sum_{k=-\infty}^{m-1} \sum_{n=-\infty}^{\infty} b_{kn}\psi_{kn}(t)$$

where

$$b_{kn} = \int_{-\infty}^{\infty} X(t)\psi_{kn}(t)dt.$$

Then we have

$$E(b_{mn}b_{kj}) = \int\int R(t-s)\psi(2^m t - n)\psi(2^k s - j)2^{m/2}2^{k/2}dt\,ds$$

$$(13.8)$$

$$= \frac{1}{2\pi}\int \hat{R}(\omega)\hat{\psi}(2^{-m}\omega)e^{-i\omega k2^{-m}}\overline{\hat{\psi}(2^{-k}\omega)}e^{i\omega j2^{-k}}2^{-m/2}2^{-k/2}d\omega.$$

For the Meyer wavelets $\hat{\psi}(\omega)$ has support on $[\pi - \epsilon, 2\pi + 2\epsilon] \bigcup [-2\pi - 2\epsilon, -\pi + \epsilon]$, and hence $\hat{\psi}(2^{-m}\omega)$ and $\hat{\psi}(2^{-k}\omega)$ have disjoint support if $|k - m| > 1$. For $m = k + 1$ we find that

$$E(b_{k+1,n}b_{kj})| \leq \frac{2^{-k+\frac{1}{2}}}{2\pi} \int_{2^{k+1}(\pi-\epsilon)}^{2^{k+1}(\pi+\epsilon)} |\hat{R}(\omega)| d\omega$$

$$\leq \|R\|_1 2^{3/2} \frac{\epsilon}{\pi}.$$

For $k = m$ we obtain the result

$$Eb_{kn}b_{kj} = \frac{2^{-k}}{2\pi} \int \hat{R}(\omega) \left| \hat{\psi}(2^{-k}\omega) \right|^2 e^{-i\omega(n-j)2^{-k}} d\omega,$$

which are just Fourier coefficients. If $\hat{R}(\omega)$ and $\hat{\psi}(\omega)$ are both C^p, then $E(b_{kn}b_{kj}) = O(|n - j|^{-p})$; we have proved

PROPOSITION 13.1
Let $X(t)$ be a wide-sense stationary process with spectral density $\hat{R}(\omega) \in C^p, p > 1$; let $\psi(t)$ be a mother wavelet of Meyer type such that $\hat{\psi}(\omega) \in C^p$; then the discrete process

$$b_{mn} = \int X(t) 2^{m/2} \psi(2^m t - n) dt$$

(i)   *is uncorrelated at scales differing by more than 1,*
(ii)   *has arbitrarily small correlation at scales differing by 1,*
(iii) *has correlation $= O(|n - j|^{-p})$ for fixed scale and translation differing by $|n - j|$.*

It should also be observed that for fixed m $\{b_{mn}\}$ is a discrete wide-sense stationary process [Ms]. These results can also be extended to certain nonstationary processes. In particular, since $\hat{\psi}(\omega) = 0$ for $|\omega| \leq \pi - \epsilon$, they can be used to analyze $1/f$ processes. These processes have spectral "densities" of the form $\hat{R}(\omega) = |\omega|^{-\lambda}$ [Fl].

The results of Proposition 13.1 still hold since the integrals are well defined. However the mean square convergence may not hold in these cases.

13.3   A series with uncorrelated coefficients

The wavelet expansions in the last section represent an approximation to the K-L expansions, but the coefficients are uncorrelated only at different scales. In this section we modify the wavelets to obtain an uncorrelated coefficient sequence. The new system will not be orthogonal but rather have an associated biorthogonal system [Z-W]. We shall also assume in this section that $\hat{R}(\omega) \neq 0$, $\omega \in \mathbb{R}$.

We first define a new function corresponding to the scaling function. It is $\theta_0(t)$ given by its Fourier transform

$$\hat{\theta}_0(\omega) := \hat{\phi}(\omega)\hat{R}^{-\frac{1}{2}}(\omega)$$

and its biorthogonal companion $\theta^0(t)$ given by

$$\hat{\theta}^0(\omega) := \hat{\phi}(\omega)\hat{R}^{\frac{1}{2}}(\omega).$$

They satisfy $\int \theta^0(t-n)\theta_0(t-m)dt = \delta_{nm}$. We define U_0 and U^0 to be respectively the closed linear span of $\{\theta_0(t-n)\}$ and of $\{\theta^0(t-n)\}$. If we define X_n by

$$X_n = \int_{-\infty}^{\infty} X(t)\theta_0(t-n)dt \qquad (13.9)$$

then we have

$$\begin{aligned}
E[X_m X_k] &= \int\int R(t-s)\theta_0(t-n)\theta_0(s-k)dt\,ds \\
&= \frac{1}{2\pi}\int \hat{R}(\omega)|\theta_0(\omega)|^2 e^{i\omega(n-k)}d\omega \\
&= \frac{1}{2\pi}\int |\hat{\phi}(\omega)|^2 e^{i\omega(n-k)}d\omega \\
&= \int \phi(t-n)\phi(t-k)dt = \delta_{nk}.
\end{aligned} \qquad (13.10)$$

The questions are: does $\{\theta_0(t-n)\}$ form a basis of U_0 and can similar modifications be made to the wavelets?

We first show that these functions induce a multiresolution-like analysis and Riesz bases for the nested subspaces. Furthermore, our wide-sense stationary process can be approximated arbitrarily closely by a K-L-like expansion whose basis is these functions.

Towards that end, define a set of scaling-function-like functions

$$\hat{\theta}_{m,n}(\omega) = \hat{R}(\omega)^{-1/2}\hat{\phi}_{m,n}(\omega), \qquad (13.11a)$$

$$\hat{\theta}^{m,n}(\omega) = \hat{R}(\omega)^{1/2}\hat{\phi}_{m,n}(\omega), \qquad (13.11b)$$

which are shifted versions of the following functions

$$\hat{\theta}_m(\omega) = \hat{R}(\omega)^{-1/2}\hat{\phi}_m(\omega),$$

$$\hat{\theta}^m(\omega) = \hat{R}(\omega)^{1/2}\hat{\phi}_m(\omega).$$

Equation (13.11a) has an interesting interpretation when $\hat{R}(\omega)$ decreases as the frequency increases (low-pass): in this case $\hat{R}^{-1/2}(\omega)$ acts as a high-pass whitening filter. The definitions in (13.11a) and (13.11b) can also be considered as a pair of linear transformations on $L^2(\mathbb{R})$, denoted by $\mathcal{R}^{-1/2}$ and $\mathcal{R}^{+1/2}$. Then, $\mathcal{R}^{-1/2}$ maps each V_m in the MRA onto a subspace U_m, the closure of the subspace spanned by $\{\theta_{m,n}(t)\}$. In fact, we have

LEMMA 13.2
For any m, $\{\theta_{m,n}(t)\} = \{\theta_m(t - 2^{-m}n)\}$ is a Riesz basis of U_m, and $\{\theta_{m,n}(t), \theta^{m,n}(t)\}$ is a biorthogonal sequence.

PROOF Without loss of generality, we prove it for the case of $m = 0$. To show that $\{\theta_0(t-n)\}$ forms a Riesz basis of U_0, we only need to show that there exist A and B such that $A \sum_n |c_n|^2 \leq \left\| \sum_n c_n \theta_0(t-n) \right\|^2 \leq B \sum_n |c_n|^2$ where $\{c_n\}$ is any sequence in ℓ^2. By Parseval's equality

$$\left\| \sum_n c_n \theta_0(t-n) \right\|^2 = \frac{1}{2\pi} \int_{-4\pi/3}^{4\pi/3} |\theta_0(\omega)|^2 \cdot \left| \sum_n c_n e^{-i\omega n} \right|^2 d\omega$$

$$= \frac{1}{2\pi} \int_{-4\pi/3}^{4\pi/3} |\hat{\phi}(\omega)|^2 \cdot [\hat{R}^{-1}(\omega)] \cdot \left| \sum_n c_n e^{-i\omega n} \right|^2 d\omega$$

$$\leq \frac{1}{2\pi} \sup_{\omega \in [-4\pi/3, 4\pi/3]} \hat{R}^{-1}(\omega) \int_{-4\pi/3}^{4\pi/3} |\hat{\phi}(\omega)|^2 \cdot \left| \sum_n c_n e^{-i\omega n} \right|^2 d\omega$$

$$\leq \frac{1}{2\pi} \sup_{\omega \in [-4\pi/3, 4\pi/3]} \hat{R}^{-1}(\omega) \int_{-2\pi}^{2\pi} |\hat{\phi}(\omega)|^2 \cdot \left| \sum_n c_n e^{-i\omega n} \right|^2 d\omega$$

$$\leq \frac{1}{2\pi} \sup_{\omega \in [-4\pi/3, 4\pi/3]} \hat{R}^{-1}(\omega) \int_{-2\pi}^{2\pi} \left| \sum_n c_n e^{-i\omega n} \right|^2 d\omega$$

$$= 2 \sup_{\omega \in [-4\pi/3, 4\pi/3]} \hat{R}^{-1}(\omega) \sum_n |c_n|^2.$$

This establishes the second inequality. Similarly, for the first inequality,

$$\left\| \sum_n c_n \theta_0(t-n) \right\|^2 = \frac{1}{2\pi} \int |\hat{\theta}_0(\omega)|^2 \cdot \left| \sum_n c_n e^{-i\omega n} \right|^2 d\omega$$

$$= \frac{1}{2\pi} \int_{-4\pi/3}^{4\pi/3} |\hat{\phi}(\omega)|^2 \cdot [\hat{R}^{-1}(\omega)] \cdot \left| \sum_n c_n e^{-i\omega n} \right|^2 d\omega$$

$$\geq \frac{1}{2\pi} \inf_{\omega \in [-4\pi/3, 4\pi/3]} \hat{R}^{-1}(\omega) \int_{-\pi}^{\pi} |\hat{\phi}(\omega)|^2 \cdot \left| \sum_n c_n e^{-i\omega n} \right|^2 d\omega$$

$$\geq \inf_{\omega \in [-4\pi/3, 4\pi/3]} \hat{R}^{-1}(\omega) \min\{\hat{\phi}(-\pi), \hat{\phi}(\pi)\} \sum_n |c_n|^2.$$

Here, we have used the fact that the minimum $\phi(\omega)$ in $[-\pi, \pi]$ is achieved at the boundary points (notice that when $\hat{\phi}(\omega)$ is symmetric, $\hat{\phi}(-\pi) = 1/\sqrt{2} = \hat{\phi}(\pi)$). $\square$

A similar result with exactly the same proof holds for $\{\theta^{mn}\}$ and $U^m = \mathcal{R}^{1/2} V_m$. While these subspaces satisfy many of the properties of the $\{V_m\}$, they are not dilation subspaces. Nonetheless we have

LEMMA 13.3
$\{U_m\}$ *(and $\{U^m\}$) form a nested sequence of subspaces in $L^2(\mathbb{R})$ such that*

$$V_{m-1} \subset U_m(U^m) \subset V_{m+1}$$

and

$$\overline{\bigcup_m U_m} = \overline{\bigcup_m U^m} = L^2(\mathbb{R}), \tag{13.12}$$

$$\bigcap_m U_m = \bigcap_m U^m = \{0\}. \tag{13.13}$$

PROOF For any $x(t) \in U_m$, the inverse of $\mathcal{R}^{-1/2}$ carries $x(t)$ to a unique $y(t) \in V_m$. But $y(t) \in V_{m+1}$, hence $x(t) \in U_{m+1}$, i.e., $\{U_m\}$ is nested. To show (13.13), we can use contradiction. Suppose there is an $x(t) \neq 0$ such that $x(t) \in \bigcap_m U_m$, then its inverse $y(t) \neq 0$ and $y(t) \in \bigcap_m V_m$, which contradicts (3.2.iv of Chapter 3). Finally, without loss of generality, (13.12) can be established if we can show

$$V_0 \subset U_1.$$

Towards that end, consider an arbitrary $x(t) \in V_0$. Then, $x(t) \in U_1$ if and only if

$$\hat{x}(\omega) = \beta(\omega)\hat{\theta}_1(\omega),$$

where $\beta(\omega)$ is a periodic function with period $2 \cdot 2\pi = 4\pi$. To show this, we notice that $\hat{x}(\omega)$ can be represented as

$$\hat{x}(\omega) = \alpha(\omega)\hat{\phi}(\omega)$$

$$= \alpha(\omega)m_0(\omega/2)\hat{\phi}(\omega/2)$$

$$= \alpha(\omega)m_0(\omega/2) \left[[\hat{R}^{1/2}(\omega)]_{[-\frac{4\pi}{3}, \frac{4\pi}{3}]} \right]_{4\pi} \left[[\hat{R}^{-1/2}(\omega)]_{[-\frac{4\pi}{3}, \frac{4\pi}{3}]} \right]_{4\pi} \hat{\phi}(\omega/2)$$

$$= \alpha(\omega) \left[[\hat{R}^{1/2}(\omega)]_{[-\frac{4\pi}{3}, \frac{4\pi}{3}]} \right]_{4\pi} m_0(\omega/2) \left[[\hat{R}^{-1/2}(\omega)]_{[-\frac{4\pi}{3}, \frac{4\pi}{3}]} \right]_{4\pi} \hat{\phi}(\omega/2)$$

where $\left[[\hat{R}^{1/2}(w)]_{\left[-\frac{4\pi}{3}, \frac{4\pi}{3} \right]} \right]_{4\pi}$ is the 4π-periodic extension of the restriction of $\hat{R}^{-1/2}(\omega)$ in $[-4\pi/3, \quad 4\pi/3]$. Since $m_0(\omega/2)$ is zero in $[-8\pi/3, \quad -4\pi/3]$ and $[4\pi/3, \quad 8\pi/3]$, and $\hat{\phi}(\omega/2)$ is zero outside $[-8\pi/3, 8\pi/3]$, the product of the last three terms can be written as

$$m_0(\omega/2) \left[[\hat{R}^{-1/2}(\omega)]_{\left[-\frac{4\pi}{3}, \frac{4\pi}{3} \right]} \right]_{4\pi} \hat{\phi}(\omega/2) = m_0(\omega/2)\hat{R}^{-1/2}(\omega)\hat{\phi}(\omega/2)$$

$$= m_0(\omega/2)\hat{\phi}_1(\omega).$$

Hence $\hat{x}(\omega)$ becomes

$$\hat{x}(\omega) = \alpha(\omega) \left[[\hat{R}^{1/2}(\omega)]_{\left[-\frac{4\pi}{3}, \frac{4\pi}{3} \right]} \right]_{4\pi} m_0(\omega/2)\hat{\phi}_1(\omega),$$

where the product of the first three factors is a periodic function with period of 4π. Hence, $V_0 \subset U_1$. The same technique can be used to show $U_0 \subset V_1$ and $V_0 \subset U^1 \subset V_2$. □

Similar to (13.11), we define the following wavelet-like functions $\gamma_{m,n}$ and $\gamma^{m,n}$

$$\hat{\gamma}_{m,n}(\omega) = \hat{R}^{-1/2}(\omega)\hat{\psi}_{m,n}(\omega), \tag{13.14a}$$

$$\hat{\gamma}^{m,n}(\omega) = \hat{R}^{1/2}(\omega)\hat{\psi}_{m,n}(\omega), \tag{13.14b}$$

based on the following functions:

$$\hat{\gamma}_m(\omega) = \hat{R}^{-1/2}(\omega)\hat{\psi}_m(\omega), \tag{3.15a}$$

$$\hat{\gamma}^m(\omega) = \hat{R}^{1/2}(\omega)\hat{\psi}_m(\omega). \tag{3.15b}$$

Then, for any given m it is not difficult to show

LEMMA 13.4
$\{\gamma_{m,n}(t)\} = \{\gamma_m(t - 2^{-m}n)\}$ *is a Riesz basis of* M_m, *where*

$$M_m = \mathcal{R}^{-1/2}W_m,$$

and $\{\gamma_{m,n}(t), \gamma^{m,n}(t)\}$ *is a biorthogonal sequence in both* m *and* n.

Together, Lemmas 13.2, 13.3, and 13.4 imply that $L^2(\mathbb{R})$ has a multiresolution-like decomposition

$$L^2(\mathbb{R}) = U_0 \bigoplus_{m \geq 0} M_m,$$

where $\bigoplus$ represents a direct sum and $\{\theta_{0,n}(t)\}$ and $\{\gamma_{m,n}(t)\}$, $m \geq 0$ form a basis (but not necessarily a Riesz basis) of $L^2(\mathbb{R})$. Hence, we have the following

THEOREM 13.5

If the spectral density of $X(t)$ is strictly positive, then it has a K-L-like expansion

$$
\begin{aligned}
X(t) &= \sum_n \langle X(t), \theta_0(t-n) \rangle \theta^0(t-n) \\
&\quad + \sum_{m \geq 0} \sum_n \langle X(t), \gamma_m(t - 2^{-m}n) > \gamma^m(t - 2^{-m}n) \\
&= \sum_n X_{0,n} \theta^0(t-n) + \sum_{m \geq 0} \sum_n X'_{m,n} \gamma^m(t - 2^{-m}n) \\
&= \sum_m \sum_n X'_{m,n} \gamma^m(t - 2^{-m}n)
\end{aligned}
\tag{13.16}
$$

and all the expansion coefficients are uncorrelated. Furthermore, the convergence in (13.16) is in the mean square sense.

PROOF The calculations of (13.10) can be used to show the coefficients are uncorrelated. Thus we need only show the mean square convergence. Consider the mean square difference between $X(t)$ and the partial sums of the expansion (13.16)

$$
\begin{aligned}
& E \left[\left(X(t) - \sum_m \sum_n X'_{m,n} \gamma^m(t - 2^{-m}n) \right)^2 \right] \\
&= E \left[X^2(t) - 2 \sum_m \sum_n X(t) X'_{m,n} \gamma^m(t - 2^{-m}n) \right. \\
&\quad \left. + \left(\sum_m \sum_n X'_{m,n} \gamma^m(t - 2^{-m}n) \right)^2 \right] \\
&= R(0) - 2 \sum_m \sum_n [\gamma^m(t - 2^{-m}n)]^2 + \sum_m \sum_n [\gamma^m(t - 2^{-m}n)]^2 \\
&= R(0) - \sum_m \sum_n [\gamma^m(t - 2^{-m}n)]^2 \geq 0.
\end{aligned}
\tag{13.17}
$$

Here we have used the fact that $X'_{m,n}$ are uncorrelated and

$$
\int R(t-s) \gamma_m(s - 2^{-m}n) ds = \gamma^m(t - 2^{-m}n).
$$

To show that (13.17) is zero in the limit (m, n go from $-\infty$ to $+\infty$), we first notice that the partial sum in the last line is convergent since it is nonnegative and bounded above by $R(0)$. Hence, the inequality in (13.17) also holds in the limit.

Furthermore, notice that $\gamma^m(t - 2^{-m}n)$ can be written as

$$
\begin{aligned}
\gamma^m(t - 2^{-m}n) &= \frac{1}{2\pi} \int \hat{\gamma}^m(\omega) e^{i\omega(t-2^{-m}n)} d\omega \\
&= \frac{1}{2\pi} \int \hat{R}^{1/2}(\omega) \hat{\psi}_m(\omega) e^{i\omega(t-2^{-m}n)} d\omega \\
&= \langle \hat{R}^{1/2}(\omega), \hat{\psi}_m(\omega) e^{i\omega(t-2^{-m}n)} \rangle
\end{aligned}
$$

and $\{\hat{\psi}_m(\omega) e^{i\omega(t-2^{-m}n)}\}$ is an orthonormal basis of all the square integrable functions in the frequency domain ($L^2(\mathbb{R})$). Hence, by Parseval's equality (for series expansions)

$$
\sum_m \sum_n [\gamma^m(t - 2^{-m}n)]^2 = \int |\hat{R}^{1/2}(\omega)|^2 d\omega = \int \hat{R}(\omega) d\omega = R(0);
$$

that is, equality holds in (13.17) in the limit.   □

13.4   Wavelets based on band limited processes

In the last section we repeatedly used the fact that $\hat{R}(\omega) \neq 0$. However many processes arising in applications are band limited and hence $\hat{R}(\omega) = 0$ for $|\omega| > \sigma$. Thus we need to modify our approach. In fact, we shall introduce a new family of wavelets based on $R(t)$ for which similar results hold [W-Z].

Accordingly let $X(t)$ be a wide-sense stationary band limited process whose spectral density $\hat{R}(\omega)$ is continuous, $\hat{R}(0) = 1$, and

$$
\hat{R}(\omega) = \begin{cases} 0 & |\omega| \geq \sigma \\ > 0 & |\omega| < \sigma \end{cases}.
$$

By changing the time scale if necessary we may take $\sigma = \pi + \epsilon$ where $0 \leq \epsilon \leq \frac{\pi}{3}$. We denote by $\hat{R}^*(\omega)$ the 2π periodic extension

$$
\hat{R}^*(\omega) := \sum_k \hat{R}(\omega + 2\pi k), \quad \omega \in \mathbb{R}.
$$

Clearly $\hat{R}^*$ is bounded, continuous, and positive.

The scaling function $\phi(t)$ is defined as that function whose Fourier transform $\hat{\phi}$ is given by

$$
\hat{\phi}(\omega) := \hat{R}^{1/2}(\omega) / (\hat{R}^*(\omega))^{1/2}. \tag{13.18}
$$

In order to qualify as an orthogonal scaling function we need only to show (Chapters 3, 9) that

$$\text{(i)} \quad \sum |\phi(\omega + 2\pi k)|^2 = 1,$$

$$\text{(ii)} \quad \hat{\phi}(\omega) = m_0\left(\frac{\omega}{2}\right)\hat{\phi}\left(\frac{\omega}{2}\right),$$

(iii) $\hat{\phi}$ is continuous at 0 and $\hat{\phi}(0) = 1$.

where $m_0\left(\frac{\omega}{2}\right)$ is a 4π periodic function in $L^2(\mathbb{R})$. The first comes directly from our definition (13.18) while the second and third may be shown by using the fact that

$$\hat{\phi}(\omega) = \begin{cases} 0 & |\omega| > \pi + \epsilon \\ 1 & |\omega| \le \pi - \epsilon \end{cases}.$$

Hence the support of $\hat{\phi}(\omega)$ is contained in the interval $[-2\pi + 2\epsilon, 2\pi - 2\epsilon]$ where $\hat{\phi}\left(\frac{\omega}{2}\right) = 1$. Thus $m_0\left(\frac{\omega}{2}\right)$ may be taken as the 4π periodic extension of $\hat{\phi}(\omega)$ since, furthermore, the $\hat{\phi}(\omega \pm 4\pi)$ are zero in the support of $\hat{\phi}\left(\frac{\omega}{2}\right)$.

The mother wavelet may then be constructed from $\phi(t)$ by the standard procedure, i.e.,

$$\hat{\psi}(\omega) := e^{-i\omega/2}\overline{m_0\left(\frac{\omega}{2} + \pi\right)}\hat{\phi}\left(\frac{\omega}{2}\right). \tag{13.19}$$

LEMMA 13.6
Let $\phi(t)$ be the function given by (13.18) and $\psi(t)$ the one given by (13.19). Then $\phi(t)$ is a scaling function that generates a multiresolution analysis $\{V_m\}$, and $\psi(t)$ is a mother wavelet such that

$$2^{m/2}\psi(2^m t - n) = \psi_{mn}(t)$$

is an orthonormal basis of $L^2(\mathbb{R})$.

PROOF It remains only to show that $\bigcup_m V_m$ is dense in $L^2(\mathbb{R})$. But this is clear since any $L^2(\mathbb{R})$ function whose Fourier transform has compact support is in V_m for m sufficiently large. Since such functions are dense in $L^2(\mathbb{R})$ so is $\bigcup_m V_m$. $\square$

The expression for the wavelet may be simplified in our case since only a finite number of terms are needed. It becomes

$$\hat{\psi}(\omega) = e^{-i\omega/2}(\hat{\phi}(\omega + 2\pi) + \hat{\phi}(\omega - 2\pi))\hat{\phi}\left(\frac{\omega}{2}\right).$$

This is the same formula as in the last section, and the wavelets are similar. From this formula, it follows that

$$|\hat{\psi}(\omega)| = \begin{cases} 1, & \pi + \epsilon \leq |\omega| \leq 2\pi - 2\epsilon \\ 0, & |\omega| \leq \pi - \epsilon \text{ or } |\omega| > 2\pi + 2\epsilon \end{cases} .$$

The results of the last section thus hold for these wavelet expansions, but now because of the compact support of $\hat{R}(\omega)$, many coefficients are zero. In fact, the wavelet coefficients of $X(t)$ satisfy

$$|E(b_{mn}b_{mk})| = \left| \frac{2^{-m}}{2\pi} \int \hat{R}(\omega) \left| \hat{\psi}(\omega 2^{-m}) \right|^2 e^{i\omega 2^{-m}(n-k)} d\omega \right|$$

$$\leq \frac{2^{-m}}{\pi} \int_{2^m(\pi-\epsilon)}^{\pi+\epsilon} \hat{R}(\omega) \left| \hat{\psi}(\omega 2^{-m}) \right|^2 d\omega$$

This is zero for $m > 0$. Hence the wavelet series is truncated, and we have the following

THEOREM 13.7
Let $X(t)$ be a wide-sense stationary band limited process such that $\hat{R}(\omega)$ has support on $[-\epsilon - \pi, \pi + \epsilon]$, let $\phi(t)$, and $\psi(t)$ be given by (13.18) and (13.19), respectively. Then

$$X(t) = \sum_n a_{1n} 2^{1/2} \phi(2t - n)$$

$$= \sum_n a_{-mn} 2^{-m/2} \phi(2^{-m}t - n) + \sum_{k=0}^{m} \sum_n b_{-kn} 2^{-k/2} \psi(2^{-k}t - n)$$

(13.20)

in the mean square sense, where the $\{a_{mn}\}$ and $\{b_{mn}\}$ are the usual coefficients of $X(t)$; the b_{kj} and b_{mn} are uncorrelated for $|k - m| > 1$ and are equal to zero for $m > 0$.

PROOF The uncorrelatedness follows from the last section. Therefore we need only prove the mean square convergence. We first observe that $R(t - s) \in V_1$ as a function of s for each t since

$$\widehat{R(t - s)}(\omega) = \hat{R}(\omega)e^{-i\omega t} = \hat{R}(\omega)e^{-i\omega t}\hat{\phi}(2^{-1}\omega)$$

$$= \sum_k \hat{R}(\omega + 4\pi k)e^{-i(\omega + 4\pi k)t}\hat{\phi}(2^{-1}\omega)$$

(13.21)

and since $\hat{R}(\omega) = 0$ for $|\omega| > \pi + \epsilon$ and $\hat{\phi}(2^{-1}\omega) = 1$ for $|\omega| \leq 2\pi - 2\epsilon$. Furthermore, since $\hat{\phi}(2^{-1}\omega) = 0$ whenever $|\omega| > 2\pi + 2\epsilon$, the periodic extension in (13.21) is valid. This tells us that $\widehat{R(t - s)}$ is a 4π periodic function times $\hat{\phi}(2^{-1}\omega)$, and hence $R(t - s) \in V_1$.

The mean square error in (13.20) is

$$
E\left[X(t) - \sum_n a_{1n}\sqrt{2}\,\phi(2t - n)\right]^2
$$

$$
= R(0) - 2\sum_n \int R(t - s)\phi(2s - n)\sqrt{2}\,ds\sqrt{2}\,\phi(2t - n)
$$

$$
+ \sum_n \sum_k \int \int R(u - s)\phi(2s - n)\sqrt{2}\,ds\,\phi(2u - k)\sqrt{2}\,du
$$

$$
2\phi(2t - n)\phi(2t - k)
$$

$$
= R(0) - 2R(0) + R(0) = 0. \qquad \square
$$

We can also imitate the procedure of the last section and introduce biorthogonal sequences $\{\theta_{mn}\}$ and $\{\theta^{mn}\}$ as well as $\{\gamma_{mn}\}$ and $\{\gamma^{mn}\}$. They are given by the same formulae (13.11) and (13.14) except that they are valid only for $m < 0$ for θ_{mn} and γ_{mn}. For $m = 0$ we may take

$$
\hat{\theta}_0(\omega) = \chi_{[-\pi-\epsilon,\pi+\epsilon]}(\omega) / [\hat{R}^*(\omega)]^{1/2}.
$$

Then the expansion coefficients

$$
X_n = \int X(t)\theta_0(t - n)dt
$$

are uncorrelated and

$$
E\left[X(t) - \sum_{n=-N}^{N} X_n\theta^0(t - n)\right]^2 = O(\epsilon) \quad \text{as } N \to \infty.
$$

The ϵ in this last statement is the one in the definiton of $\phi(t)$ (and hence of $\theta^0(t)$) and can be made small if the support of $\hat{R}(\omega)$ is close to $[-\pi, \pi]$. However, it cannot be made zero, and therefore we have only an approximation in these cases.

13.5 Nonstationary processes

Certain nonstationary processes, e.g., those with stationary increments, can be studied by methods similar to the last sections [Ms]. However there is a class of processes, the cyclostationary processes, that in general cannot but may sometimes be attacked by other wavelet methods.

A process $X(t)$ is wide-sense cyclostationary [Pa1] if

$$E(X(t+1)) = E(X(t)), \; E[X(t+1)X(s+1)] = E[X(t)X(s)].$$

An example of such a process is the modulation of a deterministic periodic signal by a random stationary process. These processes can be modeled by scaling function expansions, i.e.,

$$X(t) = \sum c_n \phi(t-n) \tag{13.22}$$

where ϕ is a scaling function, and $\{c_n\}$ is a stationary discrete process with zero mean since

$$r(t,s) = \sum_n \sum_m E(c_n c_m) \phi(t-n) \phi(s-m)$$
$$= \sum_n \sum_m \rho_{n-m} \phi(t-n) \phi(s-m)$$
$$= \sum_n \sum_m \rho_{n+1-m-1} \phi(t-1-n) \phi(s-1-m) = r(t-1, s-1).$$

If, in addition $\{c_n\}$ is assumed to be white noise, then $r(t,s) = q(t,s)$, the reproducing kernel of V_0.

However, the converse is not so easy; given $X(t)$ a cyclostationary process, is it possible to find a scaling function $\phi(t)$ such that (13.22) is satisfied?

We present one example for which it does work, the binary transmission process [Pa1]. Here $X(t)$ takes the value ± 1 in each interval $n-1 \le t < n$ with the choice of signs random. In this case $E(X(t)) = 0$ and

$$r(t,s) = \begin{cases} 1, & \text{if } n-1 \le t, s < n \quad \text{for some } n \in \mathbb{Z} \\ 0, & \text{otherwise} \end{cases}$$

Clearly $X(t)$ is nonstationary and $r(t+1, s+1) = r(t,s)$. Furthermore, $r(t,s)$ may be expressed as

$$r(t,s) = \phi(t-[s])$$

where $\phi(t)$ is the Haar scaling function (Chapter 1). But in this case

$$q(t,s) = \phi(t-[s]) = \sum_n \phi(t-n) \phi(s-n).$$

The expansion of $X(t)$ with respect to the $\{\phi(t-n)\}$ is

$$c_n = \int_{-\infty}^{\infty} X(t) \phi(t-n) dt = \int_n^{n+1} X(t) dt,$$

and hence

$$E(c_n c_m) = \int_n^{n+1} \int_m^{m+1} E(X(t)X(s)dt\,ds$$

$$= \int_n^{n+1} \int_m^{m+1} r(t,s)dt\,ds = \delta_{nm},$$

i.e., $\{c_n\}$ is uncorrelated.

The mean square convergence is easy to show as well in this case. This results in the representation of $X(t)$,

$$X(t) = \sum_n c_n \phi(t-n)$$

by an uncorrelated sequence $\{c_n\}$.

13.6 Problems

1. Let $0 < \alpha < \pi$ and let $X(t)$ be a (Gaussian) stochastic process on $[0,1]$ such that $X(0) = X(1) = 0$ whose covariance function is

$$r(t,s) = \frac{1}{\alpha \sin \alpha} \begin{cases} \sin \alpha t \sin \alpha (1-s), & 0 < t < s < 1 \\ \sin \alpha s \sin \alpha (1-t), & 0 < s < t < 1. \end{cases}$$

   (i) Convert the eigenvalue problem $R\phi = \lambda\phi$ to a differential equation with boundary conditions.

   (ii) Solve (i).

   (iii) Expand $X(t)$ in a series with uncorrelated coefficients.

2. Show the Brownian bridge is not a stationary process.

3. Let $\psi(t)$ be the Shannon wavelet, let $X(t)$ satisfy the hypothesis of Proposition 13.1. Show that the wavelet coefficients are uncorrelated at different scales.

4. Let $\phi(t)$ be the scaling function of the Shannon wavelet, let $X(t)$ be as in Problem 3, and let

$$X_n = \int_{-\infty}^{\infty} X(t)\phi(t-n)dt.$$

   Show that $\{X_n\}$ is a stationary sequence.

5. Let $X(t)$ be a wide-sense stationary process whose correlation function is $R(t) = [2\pi(t^2 + 1)]^{-1}$. Find the spectral density and use it to construct $\theta_0(t)$ and $\theta^0(t)$ of Section 13.3 for the scaling function of the Shannon wavelet.

6. Find the wavelet-like functions $\gamma_{mn}(t)$ for the example of Problem 5.

7. Let $X(t)$ be a wide-sense stationary process with $\hat{R}(w) = 1 - \frac{|w|}{\pi+\epsilon}$, $|w| < \pi + \epsilon$; 0 otherwise. Find the Fourier transform of the associated scaling function (13.8).

8. Show that the scaling function in Problem 7 satisfies the conditions needed for a multiresolution analysis.

Bibliography

[Al]   G. Alexits, *Convergence Problems of Orthogonal Series*, Oxford, New York, 1961.

[Ar]   N. Aronszajn, Theory of reproducing kernels, *Trans. Amer. Math. Soc.* 68 (1950), 337-404.

[As]   R. Askey, Orthogonal polynomials and special functions, *SIAM*, Philadelphia, 1975.

[A-W-W]   P. Auscher, G. Weiss, and M. V. Wickerhauser, Local sine and cosine bases of Coifman and Meyer and the construction of smooth wavelets, in *Wavelets—A Tutorial in Theory and Applications*, C. K. Chui (ed.) 237-256, Academic Press, New York, 1992.

[Ba]   G. Battle, A block spin construction of ondelettes Part I: Lemarié functions, *Comm. Math. Phys.* 110 (1987), 601-615.

[B-C-R]   G. Beylkin, R. Coifman, and V. Rokhlin, The fast wavelet transform, in *Wavelets and Their Applications*, M. Ruska, et al. (ed.), 181-210, Jones and Bartlett, Boston, 1991.

[B-C]   S. Bochner and K. Chandrashekharan, *Fourier Transforms*, Princeton University, Princeton, 1949.

[Br]   H. Bremermann, *Distributions, Complex Variables, and Fourier Transforms*, Addison-Wesley, Reading, MA, 1965.

[B]   P. L. Butzer, A survey of the Whittaker-Shannon sampling theorem and some of its extensions, *J. Math. Res. Exposition* 3 (1983), 185-212.

[B-S-S] P. L. Butzer, W. Splettstößer, and R. L. Stens, The sampling theorem and linear predictions in signal analysis, *Jahresber. Deutsch. Math.-Verein.* 90 (1988), 1-60.

[C-Q] C. Canuto and A. Quarteroni, Approximation results for orthogonal polynomials in Sobolev spaces, *Math. Comp.* 38 (1982), 67-86.

[Ca] J. S. Chihara, *An Introduction to Orthogonal Polynomials*, Gordon and Breach, New York, 1978.

[Cu] S. T. Chiu, Bandwidth selection for kernel density estimation, *Annals Stat.* 19 (1991), 1883-1905.

[C] C. K. Chui, *An Introduction to Wavelets*, Academic Press, New York, 1992.

[C-L] E. A. Coddington and N. Levinson, *Theory of Ordinary Differential Equations*, McGraw Hill, New York, 1955.

[C-D-F] A. Cohen, I. Daubechies, and J. C. Feaveau, Biorthogonal bases of compactly supported wavelets, *Comm. Pure Appl. Math.* 45 (1992), 485-560.

[C-M] R. R. Coifman and Y. Meyer, Remarques sur l'analyse de Fourier à fenêtre, série I, *C. R. Acad. Sci.*, Paris 312 (1991), 259-261.

[C-M-W] R. Coifman, Y. Meyer, and M. W. Wickerhauser, Size properties of wavelet packets, in *Wavelets and Their Applications*, Ruskai, *et al.* (eds.), Jones and Bartlett, Boston, 1992.

[C-H] D. Colella and C. Heil, Characterizations of scaling functions I. Continuous solution, to appear *SIAM J. Matrix Anal. Appl.* (1993).

[D1] I. Daubechies, The wavelet transform, time-frequency localization and signal analysis, *IEEE Trans. Info. Th.* 36 (1990), 961-1005.

[D2] I. Daubechies, Orthonormal bases of compactly supported wavelets, *Comm. Pure Appl. Math.* 41 (1988), 909-996.

[D] I. Daubechies, *Ten Lectures on Wavelets*, *SIAM*, Philadelphia, 1992.

[D-L-R] A. Dempster, N. M. Laird, and D. B. Rubin, Maximum likelihood from incomplete data via the EM algorithm (with discussion), *J. Royal Stat. Soc.* (B) 39 (1977), 1-38.

[E] R. E. Edwards, *Fourier Series, A Modern Introduction*, 2 Vol., Holt, Rinehart and Winston, New York, 1967.

[Fl] P. Flandrin, On the spectrum of fractional Brownian motion, *IEEE Trans. Info. Th.* 35 (1989), 197-199.

[F-R] A. Földes and P. Rèvèsz, A general method for density estimation, *Studia Sci. Math. Hungar.* 9 (1974), 443-452.

[G-S] I. M. Gelfand and G. E. Shilov, *Generalized Functions*, Vol. 1, 3, Academic Press, New York, 1964, 1967.

[G-M] A. Grossman and J. Morlet, Decomposition of Hardy functions into square integrable wavelets of constant shape, *SIAM J. Math. Anal.* 15 (1984), 723-736.

[Ha] H. F. Harmuth, *Transmission of Information by Orthogonal Functions*, Springer, New York, 1972.

[He] C. Heil, Methods of solving dilation equations, to appear in *Proc. 1991 NATO ASI on Prob. and Stoch. Methods in Anal. with Appl.* (1993).

[H-W] C. E. Heil and D. F. Walnut, Continuous and discrete wavelet transforms, *SIAM Review* 31 (1989), 628-666.

[H] J. R. Higgins, Five short stories about the cardinal series, *Bull. Amer. Math. Soc.* 12, (1985), 45-89.

[Hi] E. Hille, Introduction to the general theory of reproducing kernels, *The Rocky Mountain J. Math.* 2 (1972), 321-368.

[Jc] D. Jackson, *Fourier Series and Orthogonal Polynomials*, MAA, Oberlin, OH, 1941.

[J] A.J.E.M. Janssen, The Zak transform and sampling theorem for wavelet subspaces, preprint, 1993.

[J1] A.J.E.M. Janssen, The Smith-Barnwell condition and non-negative scaling functions, *IEEE Trans. Info. Th.* 38 (1992), 884-886.

[Ke] S. Kelly, Pointwise convergence for wavelet expansions, Ph.D. dissertation, Washington University, St. Louis, 1992.

[Ki] J. Kiefer, On large deviations of the empiric D.F. of vector chance variables and a law of the iterated logarithm, *Pacific J. of Math.* 11 (1961), 649-660.

[K-W]   T. Koornwinder and G. G. Walter, The finite continuous
        Jacobi transform and its inverse, *J. Approx. Theory* 60
        (1990), 83-100.

[K]   J. Korevaar, Distributions defined by fundamental sequen-
        ces, *Nederli Akad. Wetensch. Proc. Ser.* A 58 (1955), 368-
        389, 483-503, 663-674.

[K1]   J. Korevaar, *Mathematical Methods*, Academic Press, New
        York, 1968.

[K-T]   R. Krommal and M. Tarter, The estimation of probability
        densities and cumulatives by Fourier Series methods, *JASA*
        63 (1968), 925-952.

[L-M]   G. Lemarié and Y. Meyer, Ondeletters et bases Hibertiennes,
        Revista Matematica Iberoamericana 2 (1986), 1-18.

[L]   Y.-M. Liu, Wavelets and sampling theorems, Ph.D. thesis,
        University of Wisconsin-Milwaukee, Milwaukee, 1993.

[Lo]   S. Lojasiewicz, Sur la valeur et la limite d'une distribution
        dans un point, *Studia Math.* 16 (1957), 1-36.

[Md]   W. R. Madych, Some Elementary Properties of Multireso-
        lution Analysis of $L^2(\mathbb{R}^n)$ in *Wavlets, A Tutorial in The-
        ory and Applications*, C. K. Chui (ed.), 259-294, Academic
        Press, New York, 1992.

[Ma]   S. Mallat, Multiresolution approximation and wavelet
        orthonormal bases of $L^2(\mathbb{R})$, *Trans. AMS* 315 (1989), 69-
        88.

[Ms]   E. Masry, The wavelet transform of stochastic processes
        with stationary increments and its application to fractional
        Brownian motion, *IEEE Trans. Info. Th.* 39 (1993), 260-
        264.

[M1]   Y. Meyer, The Franklin wavelets, preprint, 1988.

[M]   Y. Meyer, *Ondelettes et opérateurs I*, Herman, Paris, 1990.

[Mo]   J. Morlet, G. Arens, I. Fourgeau, and D. Giard, Wave prop-
        agation and sampling theory, Part II, *Geophysics* 47 (1982),
        203-236.

[N-W]   M. Z. Nashed and G. G. Walter, General sampling theorems
        for functions in reproducing kernel Hilbert spaces, *Math.
        Control Signals Systems* 4 (1991), 373-412.

[O]   A. M. Olevskii, *Fourier Series With Respect to General Orthogonal Systems*, Springer, New York, 1975.

[P-W] R.E.A.C. Paley and N. Wiener, *Fourier Transforms in the Complex Domain*, Colloq. Publ. Vol. 29, Amer. Math. Soc., Providence, RI, 1934.

[Pa]   A. Papoulis, *Signal Analysis*, McGraw-Hill, New York, 1977.

[Pa1] A. Papoulis, *Probability, Random Variables and Stochastic Processes*, McGraw-Hill, New York (1991).

[P]   E. Parzen, On estimation of a probability density and mode, *Ann. Math. Statist.* 35 (1962), 1065-1076.

[P-R] B.L.S. Prakasa Rao, *Nonparametric Functional Estimation*, Academic Press, Orlando, FL, 1983.

[R-N] F. Riesz and B. Sz. Nagy, *Functional Analysis*, Ungar, New York, 1955.

[Ro]   M. Rosenblatt, *Random Processes*, Springer, New York, 1974.

[R]   W. Rudin, *Functional Analysis*, McGraw-Hill, New York, 1973.

[Sa]   G. Sansone, *Orthogonal Functions*, Interscience, New York, 1955.

[Sc]   B. Schomburg, On the approximation of the delta distribution in Sobolev spaces of negative order, *Applicable Analysis* 36 (1990), 89-93.

[S]   L. Schwartz, *Théorie des Distributions*, Hermann, Paris, (2 vols.), 1951.

[Sch] S. C. Schwartz, Estimation of a probability density by an orthogonal series, *Ann. Math. Statis.* 38 (1964), 1261-1265.

[Sh]   C. E. Shannon, Communication in the presence of noise, *Proc. IRE* 37 (1949), 10-21.

[Sp]   H. S. Shapiro, *Topics in Approximation Theory*, Springer-Verlag, Berlin-Heidelberg-New York, 1981.

[Si]   H.-T. Shim, Gibbs' phenomenon and summability in wavelet subspaces, preprint, 1993.

[Sl]   D. Slepian, Some comments on Fourier analysis, uncertainty and modeling, *SIAM Review* 28 (1983), 389-393.

[S-W]   E. M. Stein and G. Weiss, *Fourier Analysis on Euclidean Spaces*, Princeton University Press, Princeton, NJ, 1971.

[St]   G. Strang, Wavelets and dilation equations, *SIAM Review* 31 (1989), 614-627.

[St1]   G. Strang, *Introduction to Applied Mathematics*, Wellesley-Cambridge Press, Wellesley, 1986.

[So]   C. Stone, Consistent nonparametric regression, *Ann. Statis.* 5 (1977), 595-645.

[Sz]   G. Szegö, Orthogonal polynomials, Colloq. Pub., Vol. 23, *Amer. Math. Soc.*, Providence, RI, 1939.

[T]   E. C. Titchmarsh, *Eigenfunction Expansions, Parts I and II*, Oxford, London, 1962, 1958.

[V]   H. Volkmer, Distributional and square summable solutions of dilation equations, preprint, 1990.

[V-W]   H. Volkmer and G. G. Walter, Wavelets based on orthogonal basic splines, *J. Appl. Anal.* 47 (1992), 71-86.

[Wh]   G. Wahba, Interpolating spline methods for density estimation in equispaced knots, *Ann. Statist.* 3 (1975), 30-48.

[W1]   G. G. Walter, Discrete discrete wavelets, *SIAM J. Math. Anal.* 23 (1992), 1004-1014.

[W2]   G. G. Walter, Wavelets and generalized functions, in *Wavelets–a Tutorial*, C. Chui (ed.), 51-70, Academic Press, Boston, 1992.

[W3]   G. G. Walter, Pointwise convergence of wavelet expansions, to appear *J. Approx. Theory* (1994).

[W4]   G. G. Walter, Approximation of the delta function by wavelets, *J. Approx. Theory* 71 (1992), 329-343.

[W5]   G. G. Walter, Sampling theorems as part of wavelet theory, *Proc. Conf. Info. Sci. Sys.*, Johns Hopkins (1991), 907-912.

[W6]   G. G. Walter, Operators which commute with characteristic functions with applications to a lucky accident, *Complex Variables* 18 (1992), 7-12.

[W7]   G. G. Walter, Wavelet subspaces with an oversampling property, *Indag. Math.* 4 (1993), 499-507.

[W8]   G. G. Walter, Translation and dilation invariance in orthogonal wavelets, preprint, 1993.

[W9]   G. G. Walter, Negative basic spline wavelets, *J. Math. Anal. and Appl.* 177 (1993), 239-253.

[W10]   G. G. Walter, Sampling bandlimited functions of polynomial growth, *SIAM J. Math. Anal.* 19 (1988), 1198-1203.

[W11]   G. G. Walter, A sampling theorem for wavelet subspaces, *IEEE Trans. Info. Th.* 38 (1992), 881-884.

[W12]   G. G. Walter, An alternative approach to ill-posed problems, *J. Int. Equa. and Appl.* 1 (1988), 287-301.

[W13]   G. G. Walter, A class of spectral density estimators, *Ann. Inst. Stat. Math.* 32 (1980), 65-80.

[W14]   G. G. Walter, Properties of Hermite series estimation of probability density, *Ann. Stat.* 5 (1977), 1258-1264.

[W15]   G. G. Walter, Fourier series and analytic representation of distributions, *SIAM Review* 12 (1970), 272- 276.

[W16]   G. G. Walter, Hermite series as boundary values, *Trans. AMS* 218 (1976), 155-171.

[W17]   G. G. Walter, On real singularities of Legendre expansions, *Proc. AMS* 19 (1968), 1407-1412.

[W18]   G. G. Walter, Analytic representations with wavelet expansions, to appear *J. Complex Variables* (1993).

[W-B]   G. G. Walter and J. Blum, Probability density estimation using delta sequences, *Ann. Stat.* 7 (1979), 328-340.

[W-H]   G. G. Walter and G. G. Hamedani, Bayes empirical Bayes estimation for natural exponential families with quadratic variance functions, *Ann. Stat.* 19 (1991), 1191-1224.

[W-L]   G. G. Walter and Y. M. Liu, A class of bandlimited cardinal wavelets, preprint (1993).

[W-Z]   G. G. Walter and J. Zhang, Wavelets based on band-limited processes with a K-L type property, preprint (1993).

[We]   W. Wertz, *Statistical Density Estimation, A Survey*, Van den Hoeck and Ruprecht, 1978.

[X-Z]   X-G. Xia and Z. Zhang, On sampling theorems, wavelets, and wavelet transforms, to appear in *IEEE Trans. on Sig. Proc.* (1994).

[Y]   R. M. Young, *An Introduction to Non-Harmonic Fourier Series*, Academic Press, New York, 1980.

[Z-H-B]   A. Zayed, G. Hinsen, and P. L. Butzer, On Lagrange interpolation and Kramer-type sampling theorems associated with Sturm-Liouville problems, *SIAM J. Appl. Math.* 50 (1990), 893-909.

[Za]   A. Zayed, *Advances in Sampling Theory and Applications*, CRC Press, Boca Raton, 1993.

[Ze]   A. H. Zemanian, *Distribution Theory and Transform Analysis*, McGraw-Hill, New York, 1965.

[Z-W]   J. Zhang and G. G. Walter, A wavelet based K-L-like expansion for wide-sense stationary processes, to appear *IEEE Trans. Sig. Proc.* (1994).

[Z]   A. Zygmund, *Trigonometric Series*, Cambridge University Press, New York, 1957.

Index